Progress in Ape Research

Edited by

Geoffrey H. Bourne, Director

Yerkes Regional Primate Research Center
Emory University
Atlanta, Georgia

ACADEMIC PRESS INC.   New York   San Francisco   London   1977
A Subsidiary of Harcourt Brace Jovanovich, Publishers

COPYRIGHT © 1977, BY ACADEMIC PRESS, INC.
ALL RIGHTS RESERVED.
NO PART OF THIS PUBLICATION MAY BE REPRODUCED OR
TRANSMITTED IN ANY FORM OR BY ANY MEANS, ELECTRONIC
OR MECHANICAL, INCLUDING PHOTOCOPY, RECORDING, OR ANY
INFORMATION STORAGE AND RETRIEVAL SYSTEM, WITHOUT
PERMISSION IN WRITING FROM THE PUBLISHER.

ACADEMIC PRESS, INC.
111 Fifth Avenue, New York, New York 10003

United Kingdom Edition published by
ACADEMIC PRESS, INC. (LONDON) LTD.
24/28 Oval Road, London NW1

Library of Congress Cataloging in Publication Data

Main entry under title:

Progress in ape research.

Includes index.
1.   Apes—Congresses.   2. Yerkes, Robert Mearns,
1876–1956—Congresses.   3. Zoologists—United
States—Biography—Congresses.   I. Bourne, Geoffrey
Howard, Date
QL737.P96P76   599'.88   77-24746
ISBN 0-12-119350-0

PRINTED IN THE UNITED STATES OF AMERICA

Contents

Section 1

HISTORICAL BEGINNINGS OF RESEARCH ON GREAT APES

Section 2

COMMUNICATION AND LANGUAGE IN GREAT APES

Section 3

CHIMPANZEES AS BIOMEDICAL MODELS

Section 4

COMPARATIVE PERSPECTIVES OF HUMAN ORIGINS

List of Contributors

ROGER BAKEMAN, Georgia State University, and Yerkes Regional Primate Research Center, Emory University, Atlanta, Georgia 30322.

IRWIN S. BERNSTEIN, Department of Psychology, University of Georgia, Athens, and Yerkes Regional Primate Research Center, Emory University, Atlanta, Georgia 30322.

ROBERTA YERKES BLANSHARD, New Haven, Connecticut 06511.

L. D. BYRD, Yerkes Regional Primate Research Center, Emory University, Atlanta, Georgia 30322.

HAROLD J. COOLIDGE, 38 Standley Street, Beverly, Massachusetts 01915.

MEREDITH P. CRAWFORD, George Washington University, Washington, D. C. 20006.

R. K. DAVENPORT, School of Psychology, Georgia Institute of Technology and Yerkes Regional Primate Research Center, Emory University, Atlanta, Georgia 30322.

GWENDOLYN B. DOOLEY, Georgia State University and Yerkes Regional Primate Research Center, Emory University, Atlanta, Georgia 30322.

JAMES H. ELDER, Washington State University, Pullman, Washington 99163.

SUSAN M. ESSOCK, Georgia State University and Yerkes Regional Primate Research Center, Emory University, Atlanta, Georgia 30322.

ROGER S. FOUTS, Department of Psychology, University of Oklahoma, Institute of Primate Studies, Norman, Oklahoma 73069.

TIMOTHY GILL, Georgia State University, and Yerkes Regional Primate Research Center, Emory University, Atlanta, Georgia 30303.

THOMAS P. GORDON, Department of Psychology, University of Georgia, Athens, and Yerkes Regional Primate Research Center, Emory University, Atlanta, Georgia 30322.

C. E. GRAHAM, Yerkes Regional Primate Research Center, Emory University, Atlanta, Georgia 30322.

GEORGE M. HASLERUD, University of New Hampshire, Durham, New Hampshire 03824.

x   LIST OF CONTRIBUTORS

GORDON W. HEWES, Department of Anthropology, University of Colorado, Boulder, Colorado 80302.

GEORGE A. LEARY, Primate Research Center, Washington State University, Pullman, Washington 99163.

PETER MARLER, Rockefeller University, New York, New York.

RONALD D. NADLER, Yerkes Regional Primate Research Center, Emory University, Atlanta, Georgia 30322.

VINCENT NOWLIS, Department of Psychology, University of Rochester, Rochester, New York 14627.

WALTER A. PIEPER, Georgia State University, and Yerkes Primate Research Center, Emory University, Atlanta, Georgia 30322.

K. J. PRALINSKY, School of Psychology, Georgia Institute of Technology, and Yerkes Regional Primate Research Center, Emory University, Atlanta, Georgia 30322.

KARL H. PRIBRAM, Department of Psychology, Stanford University, Stanford, California.

AUSTIN RIESEN, Department of Psychology, University of California, Riverside, California 92502.

DONALD ROBBINS, Department of Psychology, Emory University, Atlanta, Georgia 30322.

DUANE M. RUMBAUGH, Georgia State University and Yerkes Regional Primate Research Center, Emory University, Atlanta, Georgia 30322.

E. SUE SAVAGE-RUMBAUGH, Georgia State University and Yerkes Regional Primate Research Center, Emory University, Atlanta, Georgia 30322.

H. F. SEIGLER, Department of Surgery and Immunology, Duke University Medical Center, Durham, North Carolina 27710.

S. D. S. SPRAGG, University of Rochester, Rochester, New York 14627.

RUSSELL H. TUTTLE, Department of Anthropology and Committee on Evolutionary Biology, University of Chicago, Chicago, Illinois 60637.

S. L. WASHBURN, University of California, Berkeley, California 94720.

BEVERLY J. WILKERSON, Georgia State University and Yerkes Regional Primate Research Center, Emory University, Atlanta, Georgia 30322.

DAVID N. YERKES, Investment Building, Washington, D. C. 20005.

FRANCIS A. YOUNG, Primate Research Center, Washington State University, Pullman, Washington 99163.

Preface

In the year 1976 when the world celebrated the two hundredth anniversary of the birth of our nation, primatologists also observed the one hundredth anniversary of the birth of Robert Mearns Yerkes.

Dr. Yerkes, born in Breadysville, Pennsylvania, on May 26, 1876, was a man of vision and persistence, who at the turn of the century, when he was only a graduate student at Harvard University, conceived of the establishment of a colony of great apes for psychobiological research. It took, however, another thirty years before he was able to fulfill his dream and open the Laboratories for Primate Biology, in Orange Park, Florida.

Dr. Yerkes had begun keeping and working with apes some years before his Orange Park laboratories were opened; his *magnum opus* "The Great Apes" was first published in August 1929, and it was "offered, in the disinterested spirit of science, to promote knowledge and enlightenment through the encouragement of honest, painstaking, unprejudiced observation."

Research with great apes was not easy to inaugurate by Dr. Yerkes. Money, accommodations for the animals, diseases, and pessimism on the part of some of his scientific colleagues all had to be overcome.

His early struggles might well have been described by Francis Bacon, "and although he was well aware how solitary an enterprise it is, and how hard a thing to win faith and credit for, nevertheless he was resolved not to abandon either it or himself; nor to be deterred from trying and entering upon that one path which is alone open to the human mind. Far better it is to make a beginning of that which may lead to something, than to engage in a perpetual struggle and pursuit in courses which have no exit." Those who have followed Dr. Yerkes as Directors of the Center have done their best to see that the work he started and developed, though "arduous and difficult in the beginning, leads out in the open country."

The staff of the Yerkes Primate Research Center considered various possibilities for celebrating the centenary of Dr. Yerkes' birth and it was decided that a conference would be most appropriate. The length and nature of the conference became an important subject for consideration, but the funds available only allowed for a two day meeting, limiting the proceedings to four sessions, and this restricted the scope and depth of the subjects to be presented.

The inclusion of an historical session was imperative on such an occasion and this offered a unique opportunity to assemble many of Dr. Yerkes own coinvestigators as well as his son and daughter, all of whom delighted us with anecdotes and accomplishments of the early years in great ape research. In turn they had the opportunity to witness the way in which Dr. Yerkes' ideas had evolved into the modern research institution that is today the Yerkes Primate Research Center of Emory University.

Following the historical session it was appropriate that the conference go on to consider contemporary great ape research, but the time available did not make possible a comprehensive presentation of this subject, and so it was decided that the remaining three sessions would deal mainly with the ongoing studies at the Yerkes Primate Research Center, complemented with presentations by outstanding scientists from other institutions. Although we regret that not every distinguished investigator in the field of great ape research was able to participate, we were especially honored to have Dr. Sherwood Washburn of the University of California at Berkeley, as chairman of, and contributor to, our final session.

In the historical introduction we were pleased to have Mr. David Yerkes and Mrs. Roberta Yerkes Blanshard open the proceedings. We are greatly indebted to them and to the co-workers of Dr. Yerkes who made the days of his directorship live for us again.

In the other sessions, we acknowledge with pleasure the thoughtful contributions of Drs. Peter Marler, Roger S. Fouts, Francis Young, and George Leary.

Over 300 participants registered for the conference and we acknowledge the many congratulatory remarks that were made publicly and privately when it was over.

Much of the success of the conference was due to the organization and dedicated work of the staff of the YPRC. Among those who should be specially mentioned are Major General George T. Duncan, Assistant Director, Mrs. Gwen Cook, and Mrs. Helen Thompson.

We are indebted to Academic Press for support in publishing the papers read at the conference and to the Director's secretary, Mrs. Gwen Cook, for carrying out the demanding task of typing the entire book in camera-ready form.

In general we feel that the conference served its purposes. These were to honor the memory of Dr. Robert Mearns Yerkes on the occasion of the 100th anniversary of his birth, to be inspired by first-hand accounts of the early work and accomplishments with great apes at Orange Park, Florida, and to emphasize some of the areas in psychobiological and medical sciences to which research with great apes is making a contribution.

GEOFFREY H. BOURNE

April 15, 1977

INTRODUCTION

Austin Riesen

University of California
Department of Psychology
Riverside, California

We are highly honored at this opening session of the
Robert M. Yerkes Birthday Centennial Conference to have with
us his daughter, Roberta Yerkes Blanshard, his son, David, and
grand-daughter, Cathie (Mrs. Hilles) Graham, daughter of David.
Roberta was for many years associated with the Yale University
Press. She and her parents gave generously of their hospitality in their New Haven home on St. Ronan Terrace, especially
to us graduate students at Thanksgiving and Christmas.
David N. Yerkes was a practicing architect in the Washington,
D. C. area. Earlier he collaborated with his father, mother,
and sister in studies of chimpanzee behavior, as he will
shortly describe to us.

In the earliest years of this century Professor Robert M.
Yerkes (1876-1956) began to discuss and to explore seriously
the problems of establishing research facilities for the study
of great apes. To understand their biology and their behavior
Yerkes saw that both field and laboratory research would be
necessary. He and his wife, Ada Waterson Yerkes, worked inten-
sively for years reviewing the widely scattered descriptions of
chimpanzees, gorillas, orangutans, and gibbons. The results of
their search and evaluations were published in a large volume
by Yale University Press in 1929: The Great Apes.

Meanwhile, the Yerkes family made a home for two young
chimpanzees. The young male, Chim, is described in Almost
Human (1925) as highly sociable and intelligent. Individual
differences among the chimpanzees of Madam Abreu's Quinta
Palatino colony near Havana, Cuba, also impressed Yerkes.
During the 1920's field expeditions were planned, financed, and
executed. Harold Coolidge, C. R. Carpenter, and Henry W. Nissen
were among the young scientists who participated in these

1

studies of wild primates.

By 1930 the Yerkes dream of a breeding and research station became a reality. Animals from Cuba, from New Haven, and directly from Africa were brought together in Orange Park, Florida. Some younger chimpanzees were installed in a penthouse laboratory suite on the roof of the Yale University School of Medicine in New Haven.

By a stroke of good fortune I had as my first zoology professor at Arizona, Dr. Elmer Darwin Ball, who had helped Dr. Yerkes search the Southeastern States during the wider geographic exploration for an appropriate site for the breeding station. After a letter from Dr. Ball I was hired as a part-time assistant in the Northern Branch of the Yale Laboratories of Psychobiology. I was simultaneously introduced to graduate studies and to a good pummeling by playful young chimpanzees in September, 1935.

The early days at Orange Park will be reviewed by our distinguished participants at this conference. In New Haven I worked more closely with Henry Nissen, as Yerkes was spending half or more of each year at Orange Park. In both branches close personal friendship between investigator and chimpanzee subject was prerequisite to the care and study of these apes. In no way did this interfere with objectivity of data, for careful controls were built into the experimental studies of sensory capabilities, learning and cooperative problem solving.

Primate veterinarians were not available in New Haven or Orange Park. Doctors practicing in Jacksonville or New Haven and medical laboratories there and at Yale cooperated fully with the staff of the Yale Laboratories. As breeding chimpanzees proceeded successfully advice from pediatricians often averted crises. Florida State Board of Health personnel cooperated in research in addition to supplying diagnostic services.

In my own studies of the development of vision I had the expert cooperation of a Jacksonville ophthalmologist, Dr. William J. Knauer. He identified early stages of optic disc pallor following visual deprivation (Figure 1). Dr. Karl Pribram was a practicing neurosurgeon in Jacksonville when he first joined in the research programs with Karl Lashley.

Mrs. Yerkes planned and planted a semi-tropical botanical garden surrounding the buildings of the Orange Park facility. Badminton and tennis courts were built and enjoyed by the staff. Nearby waters of the St. Johns River and Doctor's Lake were the scenes for fishing and sailing. Visitors arrived at the Orange Park station of the Atlantic Coast Line Railroad, as did also most equipment and supplies during the nineteen thirties and forties.

Fig. 1. Examining a young chimpanzee's eyes. William
J. Knauer, M.D., Jacksonville Ophthalmologist, at left; Austin
Riesen holding the chimpanzee, Chow, on his lap; Kao Liang
Chow holding hand and wrist for steadiness. Photograph by
K. J. Hayes.   #2400 in laboratory files.

A dream of many years that Robert M. Yerkes described to
his co-workers, but one that failed to materialize, was for
the apes to live in the forest of pine trees and oaks that
extended for many acres to the west and south of the Orange
Park buildings. Costs of building such long retaining bar-
riers were prohibitive. Instead, a compound was constructed
to surround pine trees south of the shop, kitchen, and main
animal quarters. The enclosed area was less than a full acre.
The trees did not survive climbing and branch breaking. The
space was sufficient to permit Pan plenty of room for exube-
rant displays, always a command performance when visitors
arrived. With hair bristling, feet stomping, and rhythmic
beating of hands on available surfaces, Pan often ended his
display with deft scooping of sand and gravel that was sent
flying toward the stranger(s) in his audience.

Permit me now, without further personal reminiscences and nostalgia, to introduce Mr. David Yerkes. He will describe the events preceding and following his father's first purchase of two imported chimpanzees.

HOME LIFE WITH CHIMPANZEES
Part 1.

David N. Yerkes
Investment Building
Washington, D. C.

In 1911 Robert Yerkes bought a farm in Franklin, New Hampshire, which has served as a summer place for members of the family ever since. Winters were spent in Cambridge before 1917, then in Washington until 1924, and after that in New Haven. The farm was both a summer home and a laboratory—the "field station" where our father and cooperating investigators worked on experiments with white mice, pigs, and cows, and finally, chimpanzees.

After World War 1 Father directed the Research Information Service of the National Research Council, and chaired two of the Council's committees. His work was largely administrative and he was anxious to get back into research—if possible the study of anthropoids. This had been a dream of his even before he left Harvard, and in 1923 he was thinking and talking increasingly about an anthropoid research laboratory.

The break came in the summer of 1923 with a message from Dr. William T. Hornaday, the director of the Bronx Zoo. Dr. Hornaday was calling to say that two chimpanzees were for sale in New York and that one of them was a remarkable animal—what would now be called a pygmy chimpanzee, or Pan paniscus. Father was in New Hampshire, but he arranged to buy the two chimps for $2,000. Since his total savings at that time amounted to $3,000 this took courage. However, he was not so foolhardy as to tell Mother what he had paid for the animals until long after the event.

Father supervised the remodeling of the barn to prepare it for its new tenants. At first there were two of these, but later there were four. The horse stable became a chimpanzee bedroom with Ford tires suspended from the joists of the hayloft above. Then or later, the mangers were taken out and replaced by two bed boxes, each big enough for a pair of young chimps. A partly open shed on the other side of the bar was closed in with heavy wire mesh and became a dining and play-

room.   The table was a large wooden box which was anchored to
the floor, with projecting boards on all four sides to serve
as seats.   These were intentionally made low enough so that
it was difficult to reach the table top with feet.

Early one morning Father and I drove to Concord to meet
the night train from New York.   As I remember it, the animal
dealer, Noel Lewis, got off the train with a chimpanzee
riding on each hip, but that picture is probably imaginary.
In any case, all five of us got in the Model T and drove back
to the farm in Franklin.   Mother, my sister Roberta, and Aunt
Lou Mumpoting, who had helped bring up our father and us, and
whom everyone loved, were in the yard to greet us.   When the
car stopped, Aunt Lou held out her arms and the two chimps
climbed into them as though they were coming home.

The youngsters, a male and a female, were christened
Chim and Panzee (with the accent on the zee).   I was lucky to
have them come into my life when I was going on twelve, the
perfect age to enjoy an animal like Chim as a playmate.   He
was a wonderful friend for a boy -- mischievous and full of
fun; bursting with energy, enormously curious and interested
in everything, and generally good natured.

I helped take care of the animals and assisted in experiments, but it was playing with them -- especially Chim --
that really appealed to me.   On every clear day we used to
take the two chimps up to a pasture on the hill above the
farmhouse.   There Chim -- and I -- ran wild, climbing birch
trees and riding them down to the ground, playing tag, and
generally behaving like a two-year-old chimp and an eleven-
year-old boy.   Cows were the cause of great excitement.   When
they appeared Chim would stand on the second rail of a fence,
his hair bristling all over his body, stamping his little
black feet, slapping his hands on the top rail, and making
ho-hooing noises.   Once some of us children were dancing in
the dining room to the music of a wind-up victrola.   Chim was
brought in to see the fun, and started to dance in his own
style, eyes shining with excitement, jumping up and down and
slapping his chest.   Blanche Learned, musician and co-author
with Father of Chimpanzee Intelligence and Its Vocal Expressions, said he kept perfect rhythm, in step with whoever held
his hands.

Once Chim got hold of my Brownie camera and carried it
up a tree where he examined it carefully, discovered the strap
handle, and for some time carried it around in the tree by
the handle before bringing it safely back to the ground.

I remember also how he carried things in his groin,
tucking them in between his thigh and his stomach.
Mrs. Learned described his trying to move seven stones from
one place to another in that way.   The stones kept slipping

out on one side, and he kept picking them up and putting them back, until he finally had them all where he wanted them.

Intensely curious about human speech, Chim would peer intently into your mouth when you were speaking, getting his face closer and closer, apparently trying to discover the secret of how it was done.

It is the measure of my potential as a scientist that most of my memories are of playing with the animals and not of the experiments I helped with. I do remember that we investigated the chimps' ability to discriminate between various positions and colors of boxes, and their remembrance of their findings over varying periods. Another of my duties was to go out to the barn at night and climb up to the hayloft over the chimps' sleeping room. I had bored two holes in the floor of the loft, and my job was to shine a flashlight down one hole while I looked through the other, in order to make a record of their sleeping positions.

My contributions to science were not impressive, but Father was extraordinarily patient with me. I suppose he cherished the hope, in spite of my ineptitude, that I would follow in his footsteps. This was not to be, but Father's resourcefulness and the meticulous care with which he carried out his experiments did make a lasting impression.

The association with chimpanzees during several summers was a completely delightful experience for me. When Chim died in the summer of 1924 it was a sadness for all of us; and of the other chimps who followed him, none of those I knew approached Chim in lively, good-humored intelligence. I have always felt sorry for all the boys who do not have chimpanzees as playmates.

HOME LIFE WITH CHIMPANZEES
Part 2

Roberta Yerkes Blanshard
New Haven, Connecticut

Chim measured up much more nearly to my idea of chimpanzee beauty than did Panzee -- or perhaps he set the standard for me. She was white-skinned, with large ears, and a puzzling ailment which made her lethargic and content to lie long in her bed, as well as reluctant to climb birches. He was dark-skinned, with thick, glossy, black hair and small whiskered ears; a twenty-pound well-coordinated athlete, eager to try anything, and easy to make friends with. Susan Savage of the Yerkes Laboratories (YRPRC) speaks of the pygmy chimps as being even more different from the chimpanzee than is the gorilla. And Harold Coolidge believes Chim was the first Pan paniscus to be brought to the United States. [1]

Helen Russell Pierce recalls Chim and Panzee that summer of 1923 being more free than our other chimps were later on, and walking around with everyone--with Father; with her father, Lee Russell, who photographed them; with Aunt Lou Mumpoting. Aunt Lou scolded Chim like a child for storing his porridge in his cheek. Helen's mother first saw Chim on a chilly, windy day when she was hugging a warm sweater about her. Chim came to her, took her hand and climbed up, put his arms around her neck and his legs around her waist, and nestled close against the warm sweater. She had never been used this way by a chimp before but she thought, this is a cold little chimp.

Chim was quick to find any one of the six screen doors of the farmhouse that had been left unhooked, and to clamber onto the dining table, set for lunch. But he was seldom invited to a meal there, being too eager a guest.

The farmhouse in the twenties had wood-burning stove, hand pump, icehouse in the shed, kerosene lamps and candles. Father in white coveralls put the chimps meals together in the back or summer kitchen, as separate as possible from the family's meal preparations. Their milk was sometimes a powdered form, Klim, ordered in great cans. Bananas by the stem were a staple. So was porridge or "gruel," of oatmeal or barley; and for Chim eating gruel was evidently grueling. Bananas and oranges he relished and would take from anyone, but not gruel. That he refused time and again to take from Father. He would look toward the door--this was in Washington the following winter--for someone else to come feed him, and he would accept it from anyone else: Ada, David, Roberta, Geraldine Stowell, who helped care for him and Panzee in Washington. Not dislike of me, Father concluded: "only unwillingness to accommodate me."[2] If wheatena was added to his oatmeal, Chim objected-- as I did when Father mixed old Chocolate candies into the family oatmeal.

Some of the most delightful stories about Chim, noted down as they happened, are told in Blanche Learned's portion of that little book, Chimpanzee Intelligence and Its Vocal Expressions (1925). She rejoiced in his quickness to imitate and his grace of motion.

"One afternoon," she writes, "three children were playing leap-frog in the lane, while Chim and Panzee were sitting with older people some distance away. Chim watched the children intently, for a few moments, then with a rush joined their game, leaping from back to back to the end of the line and back again without touching the ground. He liked this so well that he re-

peated it several times, while the children obligingly held
their positions". (p. 130).

"Chim was a skilled rider of the birches," she adds.
"He would climb a tree with consummate grace to the highest
point at which it would bear his weight. There he would plunge
over, carrying the top with him to a comfortable swinging distance from the ground, and would swing back and forth, or round
and round, with head down, or in any other position that might
please him, finally leaping to the ground. Then he would choose
another tree, gauging the right length of top necessary to
swing him to still another. It was a sight as beautiful in its
freedom of movement, as it was rare to our eyes, but during
those weeks it occurred almost daily, and often many times a
day" (pp.130-131).

When there were four chimps at the farm, we would lead
them up to the pasture on leashes and release them there. They
explored in pairs, an arm of one over the shoulders of the
other. Panzee, in poor health, was less active than Chim, who
explored more widely. One venture that he made by himself over
the crest of the hill brought him face to face with a dragon--
a cow. Every hair was standing on end as the routed knight
scrambled back to us.

Chim had more use for birches than just riding them.
One day--to quote Ada Yerkes--he "selected a tall birch tree
near the study, climbed to the crotch of a large limb, and
reaching out in all directions, gathered the smaller branches
and leaves back under him and lay down on them, leaning over the
edge to look down condescendingly on his excited audience below." [3] He might build three nests in one tree in an afternoon,
each higher than the last, and use each very briefly. Helen
Pierce speaks of following our fathers up to the study hilltop
with the chimps running ahead; and of seeing Chim struggle to
build a nest from two or three recalcitrant birches. Father
surveyed the hilltop full of chimp nests with a strong sense
of the incongruity of it all.

Bananas were hung out of reach to test Chim's ingenuity,
or were slipped into a fastened bottle or a tubelike box. He
failed to poke the available stick into the long box; discovered perhaps by accident how to release the bottle-with-banana.
But when he could reach the banana by swinging on a rope or
stacking boxes, Chim went at it in ways that led Father to
write: "Most surprising and impressive in Chim's behavior was
the continuity of attention, high degree of concentration on
his task, evident purposefulness of many, if not most of his
acts, his systematic survey of problematic situations, his
rapid elimination of unsuccessful acts or methods, and his
occasional pauses for reflection. I use this term without
apology, even to the behaviorist ... I have never seen a

creature give more convincing signs of reflection than does
this young chimpanzee....." [4]
Once, after he had misbehaved, climbing on the table,
screaming on the floor, and been denied breakfast until he was
quiet, Chim seemed very subdued. Mrs. Learned noted: "When
Dr. Y. turned around at the door, Chim stood up at a respectful
distance, leaned forward, and made an eloquent gesture with his
arms and hands. It seemed as though his mind and body united
in a supreme effort to make himself understood. To all appear-
ances he was trying hard to speak...." [5]
Of all the plans for housing Chim and Panzee, those involving our house in Washington concerned me most intimately.
I was then a senior in high school. It was decided--I did not
volunteer--that I should contribute my third-floor bedroom and
sleeping porch, and move into an alcove off the study on the
second floor. The cage built for the pair filled I think
about half the room. It connected with the porch by a small
swinging door, so that Chim and Panzee had free access to all
their domain. Geraldine Stowell, guests--if we still had room
for any--and I retained use of the bathroom, whose full-size
window opened onto the porch. Probably I was not the only
bather to find the presence of spectators trying, and pull down
the shade.
The house faced Rock Creek Park and was within earshot
of the Zoo. Coyotes howled in the Zoo, chimpanzees hooted on
Park Road, and Bob Mansfield, David's friend next door, per-
fected an imitation of the chimps screaming at the ashes col-
lectors. The neighbors showed surprising tolerance after it
was explained to them that the new caging was not to confine
an insane relative. Father thoroughly enjoyed answering the
doorbell with a chimp on each arm and watching the faces of
visitors.
Chim loved to have Gerry Stowell brush and rub him with
coconut oil and iodine, and when he spotted the brush in her
hand would climb onto his stool, opposite hers, ready to begin.
After he had had a workout on the sleeping porch,
swinging on the ropes, they would be tangled into knots. "He
folds the ends and loops in upon each other," Gerry commented,
"and then sits on them to hold them in place." [6] He did this
with the rug on her bedroom floor too, folding its edges up
into a heap of nest in which he would squirm and scratch his
back. Some days the porch floor would be littered with splin-
ters he had chewed from the joists. Perhaps I have protectively
forgotten the comments of the next purchaser of the house.
Tests failed to disclose what Panzee's malady was until
after her death in January of 1924. It was tuberculosis. None
of the rest of us caught it.
Chim, heartbreakingly, died of pneumonia in July, in

Cuba, where he caught the infection that was passing among
Senora Abreu's apes and monkeys. "....the heart went with him
from my anthropoid research," Father wrote that day in his
Daily Record of work for the summer. ".... I should now gladly
turn my back on it forever and think of other things."7

By July of 1925, though, there was another pair of chimps
at the farm: Billy and Dwina (Darwina), named in Scopes trial
memory for William Jennings Bryan and Charles Darwin. And by
fall there were four chimps.

The first sight of Pan, and especially of Wendy, with
her pointed devil ears, is engraved upon my memory. The third
officer of a Bull Line freighter, the Cathlamet, finally acquired in the Kamerun what he had longed for, two young chimpanzees, only to learn on reaching Boston that his family no
longer lived in the suburbs where he could keep them. An
article just out in the Boston Globe about Yerkes' experiments
with Billy and Dwina gave him an idea, and he wrote to Dr.
Yerkes.

Father and I went in the dentist's office in Cambridge
that September 15, 1925, when a telegram arrived from mother
up in Franklin about the chimpanzees for sale on a ship in
Boston harbor. Released by Dr. Eames, we made for the Cathlamet.

The ship's cargo of mahogany logs had been unloaded, but
orange mash like cooked pumpkin still greased deck and com-
panionway. In the officer's cabin we found the two chimps, the
female cowering as far under the bunk as she could get. The
price was $500. They rode to New Hampshire with us that after-
noon. This abused and frightened female, Wendy, would live to
forty-eight years in the Yerkes laboratories.

For five summers we had chimpanzee company at the farm:
Chim and Panzee in 1923; then from 1925 through 1928 Billy and
Dwina, and Pan and Wendy. Mother's farm record for 1926 begins: "At Pleasant Hill Farm: 4 Y's, 4 chimpanzees, 40 guinea
pigs, Mrs. Morford (secretary), Margaret Child (research assistant), Chauncey Louttit (investigator). No maid ..." 8
This list does not include Harold Bingham, chief collaborator
with Father in the early work in Franklin and New Haven, who
settled permanently in Sanbornton, near Franklin.

Word of our chimpanzee guests spread far and wide. Re-
reporters arrived from Boston. "Friends ... who summered
regularly in New Hampshire," Ada Yerkes wrote, "suddenly remembered us and came around to call. The Pattens (the farm
family next door) had callers from all over the state. A
familiar sight at noon on Sunday was a row of spectators in
front of the sunroom, Mr. Patten tall, thin, in blue jeans,
Mrs. Patten short and stout, guests of all ages, convulsed with
laughter over the antics of the animals who appreciated and was

always eager to put on a show. Once Dwina, the oldest, pulled
boards off the wall nearest the Pattens, and all the chimps
went over to return the many calls. We ... easily retrieved
all but Dwina, who led Robert (and I believe Mason Patten) a
chase from tree to tree, and only when she was ready came down
and was led home." 9

The first New Haven quarters for the four chimps --
Billy and Dwina, Pan and Wendy -- was a former stable, known
as the Manson Barn. It has a variety of occupants. Part of
one of the great Hillhouse Avenue properties, the barn housed
a pony and the only cow of that stately street in Philip
English's memories of his boyhood. The chimps succeeded the
cow, at I know not what interval. When the chimps moved to
grander quarters, a girls' private school called the Day-
Prospect Hill took over the barn. Yale's Astronomy Department
used it next. And in the fall of 1976 the new Graduate School
of Organization and Management absorbed the barn along with
the modern computer building next to it.

The summer of 1929, the last before the Florida laboratories opened, Father spent in strenuous travel to consult with
kindred spirits in Russia, Germany, Belgium, England, and
French Guinea in Africa. Just out of college, I got the chance
to go along, and I proved the indispensable though stammering
translator in French Guinea. We had several visits with Ivan
Petrovich Pavlov in Leningrad. We lived for seven days in the
Moscow home of Alexander and Nadie Ladygin-Kohts, talking of
her studies of chimp and child, and of the lives of scientists,
and people in general, in the Soviet Union. Father had six
sittings with the sculptor Watagin for the bust that the Yerkes
Primate Research Center now possesses.

Once outside the border of the U.S.S.R., Father summed
up his impressions -- "apart from professional interests" --
in this way: "The glass has been inverted and injustice transferred from a group called the 'workers' -- a party group of
industrial workers -- to the well-to-do professional, intellectual and industrial folk. They are being exterminated or
driven out .... Only for a few millions of the vast hord(e)
is it other than misery, misery, depression, hopelessness ...
Will communism survive here and will it upset other coun-
tries? ..." 10

July saw him conferring with the Hecks, father and son,
about the care and housing of apes in the Berlin Zoo; then
racing after Julian Huxley as he darted about the Brussels
Museum of the Congo.

In August we reached Pastoria, the collecting and research station that the Pasteur Institute in Paris maintained
at Kindia in French Guinea. Near there, for the one and only
time, Father saw chimpsnzees in the wild. Chanting Africans

propelled us on two hand cars some miles up the railroad from
Kindia to where we could see nests old and new in the trees
and, in all, saw chimpanzees. One result of that visit to
Pastoria was the arrangement for Henry Nissen to observe
chimpanzees for some weeks in the bush, and then to bring back
sixteen chimps to the new anthropoid station in Orange Park.
      The Great Apes was published by the Yale University
Press that memorable August of 1929; we saw the first copy at
Marseilles, on the way home from Africa. Yale approved plans
for the Florida laboratory in December and the Yale Anthropoid
Station opened in June of 1930.

REFERENCES

1. Susan Savage and Harold J. Coolidge speaking at the Yerkes
        Centennial Conference, Atlanta, 10/26/76.
2. Yale Medical Library, Yerkes papers, Supplementary File,
        Drawer 3. "Accounts, Cuba, Record Book 2, 1924",
        pp. 1-5.
3. Ada W. Yerkes, "Family Matters," p. 385 (typescript bound
        in 3 vols. to be archived at Yale).
4. Yerkes and Learned, Chimpanzee Intelligence and Its Vocal
        Expressions, pp. 48-49.
5. Ibid., p. 93.
6. Yerkes Papers, Sup. File, Drawer 3. Stowell, "Record Book
        2," p.23, March 14, 1924.
7. Loc. cit., Drawer 9; with diaries.
8. A. W. Yerkes, "Family Matters," p. 388.
9. Ibid., p. 392.
10. Yerkes Papers, Sup. File, Drawer 9. "Log and Notes of
        European-African Journey, 1929, June-October"; pages
        under F.

FIVE YEARS IN NEW HAVEN AND ORANGE PARK:
AN APPRECIATION

Meredith P. Crawford
The George Washington University

Thirty-eight years should provide enough time in which to gain some perspective. When Mrs. Crawford and I left Orange Park in the summer of 1939, I believe we went with a feeling of satisfaction, of time well spent, and of valuable experience gained. At the time, like most young men, I was eager to move on to something else - in my case to college teaching. While I thought it was a good experience then, that impression has been amplified and enhanced over the almost four intervening decades.

In my allotted twenty minutes, I hope to accomplish two things. First, I will join my colleagues of the 'thirties in saying something about our lives in those days as researchers in the Yale Laboratories of Primate Biology. Second, I will adopt a larger perspective and touch on some of Dr. Yerkes' major accomplishments, both within and beyond the laboratories, which were endeavors of national scope. I will attempt this latter because my particular five years have been illuminated for me, in retrospect, as I have come to recognize something of the wealth of experience and the broad perspective which he brought to our work with the laboratory animal of his choice. As a young man, I worked for and with a man whose perspective and point of view I could not fully realize at the time. Since then, I have had opportunities to deal with some of the problems which he wrestled and to see the fruition of some of the ideational seeds he planted.

First, then, I will reminisce a bit about New Haven and Orange Park between 1934 and 1939. My first full-time employment was in the laboratory at New Haven, with a salary of six-hundred dollars a year and a room in the complex of the New Haven Hospital. That was a welcomed opportunity in those Depression days. We lived simply - Shirley Spragg and I cooked our breakfasts in one of the laboratory kitchens. I had come from Columbia with a stack of raw data sheets from which to write a Columbia Ph.D. thesis.

My assignment for the first year was to write that thesis

and to work with Henry Nissen on food-sharing between pairs of young chimpanzees. The environment in which to accomplish this was most favorable.

The relevant scientific community centered around 333 Cedar Street, the common address of the Institute for Human Relations and the Medical School. It was a stimulating one. There were many able graduate students in Psychology and Physiology, among them Neal Miller, Carl Hoveland and Theodore Ruch, as well as some of the Yerkes students here present. The younger research and instructional staff included such people as Arthur Melton, Leonard Dobb, Donald Marquis, and Robert Sears, as well as Nissen and Carlyle Jacobsen of the Laboratory staff. Among the senior professional group were, in addition to Dr. Yerkes, Clark Hull, Walter Miles, and Mark May, the Director of the Institute.

The arrangements for learning both formal and informal, were many and inviting. There was Yerkes' seminar in comparative psychology and psychobiology and Hull's seminar in learning. There were luncheon groups at the Institute "Blue Room" and afternoon teas in the lounge of the Medical School, in both of which extended discussions among faculty and students took place. There was a Yale Hall of Graduate Studies where wider university contacts were available during the evening meal – and later, for me, in residence. As you can imagine, it was a great place in which to begin a career.

Amid this stimulating intellectual and social environment fell the routine duties of the younger research staff of the Laboratory. We fed the young chimps their evening meals at their dining tables and we struggled to acquire the art of firm discipline. We cooked the rice and wheat germ which made up part of their staple diet. We weighed and measured them at regular intervals, and we arried them to and from experiments on our backs. We collected fresh stool samples in the early morning for examination in the hospital pathology lab. Through these several chores we were early indoctrinated into the whole animal, the psychobiological point of view. Sometimes we psychologists thought it was more biological than psychological!

I soon began to participate in the scientific work of the Laboratory, involving the planning of research projects and a critique of plans by Yerkes and Nissen. Apparatus was fabricated in the shop, experiments were accomplished, moving pictures were made of the animals at work, results were presented in seminars, and reports were written. Things moved apace.

During these years there was a great deal of interest in the work of the Laboratory both in scientific circles and among people generally. To meet the former, we young fellows were encouraged to give papers at scientific meetings and some

of our results were picked up in textbooks not long after.
Publications was a must. Concerning the public media, Dr.
Yerkes steered a careful course, on the one hand between accurate and scientific, yet intelligible writing for the public
and sensationalism on the other. I recall with vividness
Dr. Yerkes dismay over a lengthy article in a New Haven news-
paper under the title "Chimps no Chumps; Champs at Begging -
Generous Too", which recounted some of the work Nissen and I
were doing. These words alone hardly suggested the scientific
objectivity toward which we strove!

Toward the end of my second year I migrated to Orange
Park in my second-hand, one-hundred dollar Model A Ford Road-
ster. (I was making twelve-hundred dollars a year then). There
the scene was different. Our predominantly outdoor life
centered around the open-air cages and experimental equipment
of the Lab. We dealt most indirectly with the large animals
and most directly with the infants. My first duty was the care
and feeding of the baby Pete, who had been separated from his
mother shortly before. I gave him daily baths with Palmolive
soap and mixed his formula. I learned to put a nipple on a
wide-mouthed bottle some time before it was a requirement in
my own household!   Again, actual practice in the psychobiological point of view.

Life was pleasant in Orange Park, but in quite a different way than in New Haven. In terms of research, we learned
to know our subjects more intimately and we observed the full
range of the life cycle. Intellectual work was still the
order of the day. We had the journals and other resources,
even including an early version of microfilm for literature
review. Formal seminars continued, especially during Dr.
Yerkes' six-month residence each year.

We also had the amenities of small-town life near an arm
of the sea. Mrs. Crawford and I spent two years as the tenants
of a ten-room, furnished house on a five-acre lot, complete
with resident yard-man and a library of Robert Burns in the
study - all for twenty-five dollars a month! We owned an auto-
mobile and a sailboat and bought a bond a month. On Saturday
afternoons we sailed and went to Jacksonville in the evening
for shopping and the movies. Jim Elder and I sawed and chopped
all our firewood from fallen trees and branches on the square
mile of Laboratory property. Kenneth Spence and I played some
first-class tennis on the Laboratory court. The whole group
was often entertained in the home of Dr. and Mrs. Yerkes, as
was often the case in New Haven.

We enjoyed the visits of the scientific advisory board
that came to the Laboratory each year and our opportunities to
talk with them about our work. They critiqued our ideas, gave
us some new ones. They also pointed out to us the uniqueness

might be realized if its scientists and practitioners were united in a single organization. During 1941 and 1942 he worked with his several colleagues on the development of an idea which culminated in the Inter Society Constitutional Convention of Psychologists that was held in New York City in May of 1943. The issue under debate was the formation of a loose federation of existing groups of psychologists versus the establishment of a new, integrated organization.

Dr. Yerkes worked toward the latter objective. He witnessed the realization of his objective in 1945, when the Council of the American Psychological Association adopted a new set of By-Laws. Through its divisional structure this new constitution provided places for the several diverse groups whose members were to be represented in a newly constituted Council of Representatives. A Central Office for the new APA was established in Washington with a full-time staff. (It is pleasant to note that Mrs. Helen Morford, whom we knew as Dr. Yerkes' charming and efficient secretary for many years, served in a senior staff capacity in that Central Office in its early years.) The approximately six thousand psychologists of the time were finally united under a common standard. The organizational pattern established at that time has proved its flexibility over the years in accomodating the interests of its current forty-one thousand members. Dr. Yerkes' long-time objective is well summarized in the opening clause of the current APA By-Laws. It states the objective of the Association is "...to advance Psychology as a science and profession and as a means of promoting human welfare."

In summary, I have tried to say that my five years in New Haven and in Orange Park with Dr. Yerkes and my old colleagues were a valuable learning experience and provided a fortunate beginning for a professional career and family life. That, in itself, is enough for which to express appreciation. When those years are reviewed and re-thought in the perspective of Dr. Yerkes' three major accomplishments that I have alluded to, they take on a larger significance for me.

REMINISCENCES OF THE YERKES ORANGE PARK LABORATORY IN THE MID-1930s

George M. Haslerud

University of New Hampshire
Durham, New Hampshire

As a student at the University of Minnesota I had read some of the articles and books by Dr. Yerkes and knew he had started the Psychology Department there and appointed its first staff though he was never in residence, but only when I arrived at the Yale Institute of Human Relations did I get to know him personally. It was Carlyle Jacobsen who offered me in 1934 an opportunity to join at the Yale Laboratory of Primate Biology his neurological and behavioral researches on delayed reaction in rhesus monkeys. "Of course, there isn't any pay", he said, "but the problems are interesting ones". I recalled the story of Mark Twain instructing a Nevada miner how to secure a job when none is available--start working hard for nothing.

I was glad I did because the Institute of Human Relations had probably in 1934-35 the most remarkable collection of faculty and post-doctorate fellows in the United States. About a dozen later became presidents of the American Psychological Association. The very air was electric with ideas and intellectual stimulation at the inexpensive lunches in the Institute's Blue Room. Also each week the Institute's colloquim sampled contributions from Dr. Yerkes' domain in Psychobiology and from those of Miles, Dodge, Mark May, Hull, and from the medical school ones of Fulton and others. Monday nights everyone went to Clark Hull's seminar.

Soon after my arrival in New Haven Jacobsen introduced me to Dr. Yerkes. He seemed austere and business-like as he outlined the conditions and precautions for work at the laboratory. During later weeks I was called to his office several times to receive reprints and autographed books, e.g., The Great Apes. Each time he became more approachable. He took an interest in the frontal lobe studies with monkeys on delayed response and motor restitution on which I was assisting Jacobsen. Then after several months came the great day for me when he announced at the conclusion of such a conference

that Yale and his budget had found funds to pay me $150. a
month and invited me to go in the spring to the Orange Park,
Florida part of the laboratory to work with chimpanzees. Already I had taken my turns for weekend care of the colony of
chimpanzee children in the penthouse of the medical school and
had begun an exploratory cognitive study with Nissen, but, as
Dr. Yerkes portrayed it, the Orange Park laboratory would open
opportunities for some personal research with both chimpanzee
children and adults.

When I arrived at Orange Park in my second-hand $100.
Model A Ford, I suffered considerable cultural and physical
shock. The drinking and shower water were heavily impregna-
ted with hydrogen sulfide. Almost at once I began to have
skin infections and trouble with the Florida insects. I was
reminded that I had exchanged Minnesota and New England for
the sub-tropics by the evening boom of alligator bulls in the
backwaters of the St. Johns river. The first nights my car
nearly collided with multitudes of scrawny cattle bedded down
on the highways before I came sufficiently wary to survive
under Florida's open-range law. The two-mile unimproved road
from the village to the laboratory went through an exotic
region of palmetto and pine with scattered unpainted shacks of
the black turpentiners, sprawled on narrow front porches sur-
rounded by their many dogs. The chain-link fence enclosing
the laboratory grounds had a big gate over a bridge of widely
spaced pipes to keep out the almost wild razorback hogs. Mr.
Atwater, the genial and resourceful superintendent of the
Orange Park laboratory, directed the work of feeding and car-
ing for the animals and helped the researchers to plan and
make apparatus. Mrs. Parmenter, a prim old lady of the Old
South who both resented and appreciated Yankees, agreed to
provide me with room and board as she had done for a number of
the others at the laboratory in previous years. But it was
not only good food one got at the Parmenter table; there were
also endless bits of natural and historical lore and curious
tales about persons in the village and at the laboratory.

I found that adult chimpanzees differed a great deal
from the immature ones I had known at the New Haven laboratory. The flagrant genital rosettes of the females at the
middle of their cycle were carefully measured and recorded by
Drs. Yerkes and Elder as they carried on their program of sex
research on the chimpanzee. Many of the adults liked to play
tricks on each other and on humans too. A favorite was to
fill their mouths with water from the fountain and then cling
innocent-like to the chain fence until a passerby could be
squirted. Especially interesting to me was the surgical skill
of a big male. Dr. Yerkes suggested that I let Jack remove a
thorn that had broken in one of my fingers. When I held it
up to the fence and Dr. Yerkes pointed to the thorn, Jack be-

came all attention as he gently squeezed the injured area with big forefingers and sucked with pointed lips. The thorn came out and I suffered no after-infection. Dr. Yerkes knew all the animals so well that his personality sketches made each an individual with a name and distinctive characteristics.

Hardly had I received minimal orientation to the laboratory before I was assigned the daily care of week-old Peter whose mother had rejected him. He was to be reared in isolation from his kind. His monthly examinations by a Jacksonville pediatrician (Dr. Yerkes wanted no veterinarian for Peter) prepared me for my own children's same sequence but with double the time for appearance of various behaviors. The floor of Peter's bassinet-cage was covered daily with new white paper toweling for sanitary purposes. It wasn't long before I noticed and on the weekly report which we all wrote on yellow scratch pads informed Mrs. Morford, Dr. Yerkes' very efficient and cordial secretary, that Peter constantly clutched a piece of paper and resisted having it taken away. I found I could remove all except a square centimeter scrap without starting a tantrum. Tom McCulloch followed up this observation with several published papers where he used paper toweling as a reward in delayed response and discrimination studies. At 17 months Peter could delay 60 minutes with 100% accuracy, much longer than had ever been reported with food as a reward. At the University of Wisconsin Harry Harlow began a new line of research on mother-child relations in rhesus monkeys. For the origin of his techniques he gave the following in his 1958 APA presidential address: "The (rhesus) infants on cloth mothers clung to those pads and engaged in violent temper tantrums when the pads were removed...such contact-need or responsiveness had been reported previously by Gertrude van Wagenen for monkeys and by Thomas McCulloch and George Haslerud for the chimpanzee".

The way I got started on my study of "The Effect of Movement of Stimulus Objects upon Avoidance Reactions in Chimpanzees" illustrates how Dr. Yerkes led the research projects of the laboratory. He and Mrs. Yerkes had tested avoidance in 29 young and mature members of the colony by rating +4 to -4 their responses to five objects: shuttlecock, rubber tubing, a small rubber dog, and live tortoise and glass snake. Despite not knowing the individual experience and species traditions concerning these objects and also their inability to regulate the activity of the tortoise and snake, they tentatively proposed "To proceed from the present point of exploratory survey to the experimental evaluation of movement as factor in apparently hereditary avoidance response". Dr. Yerkes asked if I would be interested. The isolation of Peter from his kind and complete history of his environment

would provide a critical test; also the avoidance behavior
might be measured more objectively than by ratings. Dr. Yerkes
imposed no certain method, stimulus objects, or plan. Rather
he made constructive criticisms of what I suggested from his
knowledge of chimpanzee behavior in general and of their reasons in the exploratory study he and Mrs. Yerkes had just completed in particular. To control movement he seemed pleased
with my idea of mounting in an aggressive pose one of each of
similar pairs of tortoises, snakes, and small alligators and
of restraining by leashes the live animals in the pairs. Fire
from the wick of a kerosene lamp and a pictured flame on a
similar lamp were also used. Measurement of the time and
mount of avoidance came from use of Kurt Lewin's concept of
valence. For example, would the animal reach for a piece of
fruit 90 cm. from a potentially feared object more than when
it was at 10 cm. and both less than when the food could be
reached with no feared object present at all? The actual behavior was recorded for each trial and later the descriptions
were sampled and rated by six psychologists familiar with
chimpanzee behavior on a scale from 0 to 4. Statistical tests
indicated significant rapid adaptation of the young chimpanzees but not of the adults. Moreover, the chimpanzee children
more quickly lost their fear of the inanimate forms than of
the animate kinds of stimuli. Also, as had been reported by
Dr. and Mrs. Yerkes there were wide individual differences
among the animals on which objects were feared most.

The crucial testing of Peter with the same technique and
objects was done at 7 and 15 months. At 7 months the affective disturbance most interpretable as avoidance occurred primarily to moving objects, but there was also a little aggressiveness toward certain objects. At 15 months no differenti-
ation on the basis of movement was discernible, but a more
intense avoidance of a wider range of objects existed along
with exhibition of much more aggression toward some of these.
Peter at 15 months had come to behave like a socialized chimpanzee that had had previous experience with both the animate
and inanimate objects.

At Yale in the Psychology Department but also in the
Institute of Human Relations the frustration-aggression-hypo-
thesis was much discussed in the 1930s. I had been intrigued
by two kinds of behavior in the avoidance study. My six
raters had difficulty placing it on the 0 to 4 scale but settled on a consensus at 2 1/2. Reading from the final scale:
"Retreats to safe distance for bluffing behavior, shakes tire,
pounds floor or wire netting of cage wall, moves hands as
though to push away stimulus object. Or-- retreats to safe
distance, avoids looking toward stimulus object, plays abstr-
actedly with tiny objects in the cage or with parts of its
body or diverts itself by somersaults". I decided to see if

such opposite behaviors would occur in a frustrating situation where there was no feared object. Also of interest would be comparison of chimpanzee children and adults for frustration tolerance.

I used a variety of the pull-in techniques to build up an expectation of food reward and then tested the effect of three frustrating conditions repeated up to eight times. The two wheels on which an endless V-type belt was mounted were concealed respectively behind a curtain and under a platform above which the animal extended his arm through a hole in the fence. The unit task was a pull of 90 cm. (4 turns of Veeder counter); during the food or standard trials this brought a piece of orange in a tin container clipped on the belt from behind the curtain. A falling door prevented access to the belt between trials. After the expectation had been built to a criterion level, I frustrated the animals by an empty food tray after the standard pull, or by interrupting the pull when the tray loaded with orange came in sight by dropping the door or thirdly by letting the animal continue to pull without even an empty tray until it withdrew its arm from the hole. The results of these experiments indicated that frustration is much more a function of how a reward-expectancy is thwarted than of the simple loss of reward. The interrupted pull led to the most pronounced frustration, with the unlimited pull second, and the empty container last, roughly respectively in the proportion of 2, 4/3, and 1 times the frustration manifested to the empty food container. The child chimpanzees on the whole were much less resistant to frustration than the adults, though several adults had scores not much above those of the most resistant children. Both young and adult animals when greatly frustrated exhibited the two types of behavior mentioned at the 2 1/2 rating level on the avoidance scale. Some tended to be extrapunitive, working out their tension on the environment by beating the door closing the hole in the fence, screaming, running around the cage. Others trembling slightly retreated to the back of the cage, doodled with finger writing in the dust on the floor, scratched or lightly bit themselves. These contrasting behaviors are analogous to those instances reported by the Yale psychologists and sociologists in their book, Frustration and Aggression. Apparently aggression can be turned either outward or toward the self in both men and chimpanzees. Moreover, frustrational and fear (avoidance) states seem to have much in common.

I ought to recall also the cooperation and mutual stimulation among the researchers at the Orange Park laboratory, the many good times socially at their homes and especially the evenings of hospitality and games at the Yerkes home on the St. Johns river, the visits to the laboratory almost every week of distinguished scientists, but I want to use my remaining minutes to recount a dramatic personal experience that indicates how Dr. Yerkes' wisdom extended to understanding human as well as chimpanzee behavior. When in late August of 1935 I drove to Ann Arbor, Michigan for the annual meetings of the American Psychological Association, I went through the TVA region in Tennessee and was so impressed with plans for salvaging that "poorhouse of the nation" that when offered a position in the Psychology Department at the University of Tennessee I impulsively accepted and wrote Dr. Yerkes a letter of resignation. Angry and disappointed he denounced my irresponsible decision but then ended with good wishes for my success in the academic position and, best of all, with an offer to spend my 3-month vacations at the Orange Park laboratory. I gratefully made use of that opportunity for research the next three summers.

Even after I was no longer at the laboratory, Dr. Yerkes and I continued to correspond and exchange home visits, especially after I located at the University of New Hampshire. This is not far from Franklin, New Hampshire where Dr. Yerkes since the beginnings of the century had a summer farm which continue in his family's possession. The Franklin Farm still has the old barn containing the living quarters and the 1920s apparatus from the first psychological experiments on chimpanzees in the United States. Behind the barn are the descendants of the birches at whose tops pioneer Chim made chimpanzee nests. Also at the brow of the hill is the study cabin where Dr. Yerkes did a great deal of his writing and its porch overlooking a distant lake and mountain where he dreamed and planned for the scientific study of chimpanzee and man.

REFERENCES

Dollard, J. et al., Frustration and Aggression, New Haven; Yale Press, 1939.

Harlow, H. F. The Nature of Love. Amer. Psychologist, 1958, 13, 673-685.

Haslerud, G. M. The effect of movement of stimulus objects upon avoidance reactions in chimpanzees. J. of Comp. Psychology, 1938, 25, 507-528.

Haslerud, G. M. Some interrelations of behavioral measures of frustration in chimpanzees. Part III of symposium on "Frustration as an Experimental Problem." Character and Personality, 1938, 7, 136-139.

Jacobsen, C. F., Taylor, F. V., and Haslerud, G. M. Restitution of function after cortical injury in monkeys. American Journal of Physiology, 1936, 116, 85-86.

McCulloch, T. M.   The role of clasping activity in adaptive
        behavior of the infant chimpanzee.   I. Delayed response.
        J. of Psychology, 1939, 7, 283-292.
McCulloch, T. M.   The role of clasping activity in adaptive
        behavior of the infant chimpanzee.   II. Visual discri-
        mination.   J. of Psychology, 1939, 7, 293-304.
McCulloch, T. M.   Affective responses of an infant chimpanzee
        reared in isolation from its kind.   J. of Comparative
        Psychology, 1939, 8, 437-445.
Yerkes, R. M. and Yerkes, A. W.   Nature and conditions of
        avoidance (fear) response in chimpanzees.   J. of Compar-
        ative Psychology, 1936, 21, 53-66.

ROBERT M. YERKES AND MEMORIES OF EARLY
DAYS IN THE LABORATORIES

James H. Elder

Washington State University
Pullman, Washington

It was in the depression year of 1930 that I had the
good fortune of receiving a research assistantship with Dr.
Yerkes in the Laboratories of Comparative Psychobiology. An
alternative was public school teaching in Wyoming where the
pay at the time was relatively good and we needed it. But the
Yale assistantship with a $1000 stipend was exceptional when
compared with the prevailing national range of less than $500.

As an undergraduate at the University of Colorado working with Karl Muenzinger, and to some extent with L. C. Cole,
who was Dr. Yerkes' first Ph.D. student at Harvard, it was
pressed upon me through my mentors and my readings that, whatever sins I might choose to indulge in, I should by all means
shun anthropomorphic language or thought.

The attractive prose of Aristotle, Montaigne, Darwin and
Romanes in their descriptions of animal behavior should be put
aside. Beer, Bethe, von Uexkull, Lloyd Morgan, Jaques Loeb,
and Watson stand out as sources of influence in my early years
in psychology. As a result of this indoctrination I scrupulously avoided putting thoughts into the head of my rat. It
was with this background that I came to work in the New Haven
Laboratory in September 1930.

It may be possible to avoid anthropomorphic descriptions
when working with chimpanzees, but it is very difficult and
eventually one weakens. I must credit Doctor Yerkes with
breaking down my rigidity on this point, although it took him
some time to do it. (Of course the animals helped greatly in
the process). I do not recall that we spent any time debating
the matter except on one occasion a few years later when I
thought he had gone a bit too far. We were working over the
joint monograph on "Oestrus, receptivity, and mating in
chimpanzees".

The language to which I objected was (in describing the
mating situation)" ....when the facts warrant it, the male

usually is expectant of copulation, takes his preferred position in the cage and awaits the approach of the female, with
evident assurance. Rarely is he mistaken in his diagnosis of
the condition." In another situation the male "inspects her
and behaves in accordance with his findings".

Perhaps it was a trace of my early training which gave
me trouble with such phrases as "when the facts warrant it",
"with evident assurance", "diagnosis of the condition", "in
accordance with his findings". I must now admit that these
brief quotations, even taken out of context as they are,
should give the reader a much more accurate picture of what
was going on in the pen than Jaques Loeb could possibly have
done.

Even though we did not discuss my problem openly it is
clear that Doctor Yerkes was aware of it, as seen in a paragraph from his letter of May 20, 1936, commenting on a joint
paper we were preparing: "I have tried to write freely esca-
ping inhibitions so far as possible and taking certain risks
and liberties which may make you shudder. I haven't apologized for the use of subjective terms, where objective description would have extended a sentence into a paragraph or a
paragraph into a page".

Of course we have long since cast away our inhibitions
about anthropomorphic language. Many recent best sellers
about animals would never have made it otherwise, but even the
scientific literature is quite different in this respect.

The intense concern around the turn of the century (plus
or minus 20 years) on the question of mind in animals may have
distracted us from attending to more important problems. But
as we now look back on the relatively short and recent history
of comparative psychology it is clear that, after Darwin,
passing through this phase was necessary and inevitable. Did
we ever answer the pressing questions of that time? I think
not. Having exhausted ourselves in debate we went our separate ways with our tentative answers. (However, in recent
publications, Don Griffin is raising these questions again,
perhaps on a more sophisticated level!)

As Doctor Yerkes has described in several historical
publications, his basic proposal for psychobiological research facilities, presented first in 1916, was not substantially funded until 1929. However, interim support permitted
the establishment of the Primate Laboratory in New Haven.
When I arrived the laboratory occupied an old but quite elaborate carriage house on Prospect Street - 135 I think it was.
The structure provided space for offices, a large kitchen,
some experimental rooms and adequate quarters for our animals.
Research personnel with offices in the Prospect laboratory
were: Dr. Henry W. Nissen, Dr. Louis W. Gellerman, NRC fellow,

Kenneth W. Spence and myself, as graduate assistants. The
animal custodian at that time was Mike Figurniak, a meticulous and most faithful person who seemed to love his job and
learned to speak chimpanzee some time before the rest of us
had mastered the language. Incidentally, I never debated
anthropomorphism with Mike, either.

Chimpanzee residents at this time were Kani, Song, Moos,
Kindia, Boy and two infants, Kambi and Bimba. Bokar joined us
a bit later. All of these had been collected by Dr. Nissen
only a few months earlier.

In addition to sharing with Spence the chores of food
preparation and serving meals to the older group, I had a
special responsibility of feeding the infants. Their diet
consisted chiefly of Pablum, mashed banana, and a mixture of
milk and graham crackers. This assignment was most delightful and rewarding - even though the babies seemed to be
learning faster than I as to how this routine should be
handled. In reflecting on my observations of these rapidly
developing little creatures I cannot help wondering if we
might not have avoided all of our controversy over mind in
animals, if Descartes at an early period in his life, had
been given the care and feeding of two baby apes.

Although others were soon to join us after we moved in
the summer of 1931 into the spacious quarters of the Institute of Human Relations, Spence and I were the only research
assistants. As such we were not only required to feed and
care for animals but expected to devote a substantial part of
our time to research.

Doctor Yerkes wasted little time setting the two of us
to work on studying the primary sense modes of vision and
hearing. We were not assigned these areas. There was not
coercion except of a subtle sort which made it clear to us
that basic sensory research had high priority. Work in experimental psychology as an undergraduate made the investigation of hearing an easy choice for me. Besides Doctor
Yerkes had referred to the area of hearing in chimpanzees as
a "terra incognita". It is quite exciting to work in such
places as that. I like to think also that it was Spence's
choice to study vision.

In the early 1930's we could not have wished for a more
encouraging environment for research. We did get off to a
false start on the auditory research, quite clearly for the
reason that we were under the strong influence of Pavlov. The
result was a massive custom-built chair in which we planned
to restrain chimpanzees and subject them to tones and electric shock. It was natural that we should dub this monstro-
sity the "chimchair". With its thick, 4-inch leather straps
and cast aluminum device for holding the necks of our sub-

jects in place, a comparison with the electric chair is not inappropriate.

By methods quite similar to current behavior modification practices two animals were persuaded to enter the room, sit in the chair and allow themselves to be strapped in. And then Kani told us with dramatic emphasis that she was finished with whatever trickery we were up to by suddenly swiveling in the chair with the neck piece lightly closed, a movement which certainly would have broken an ordinary neck. Kani was tough and also highly motivated. That evening she put an end to our false start. We had a chimchair and the waste of six months of precious time, but we learned so much during the period that we were actually congratulating ourselves. I had by this time become sufficiently acquainted with chimpanzees to believe that we could get measures of auditory acuity more readily by treating them like human children. Thus evolved a testing situation not unlike that used in audiometric clinics. Thenceforth progress was orderly and rapid.

An advantage of having captive subjects is that they can be observed continuously. Obviously aware of the exceptional opportunities we had, Doctor Yerkes pressed upon us the need for keeping daily records of behavioral and other events not related in any way with formal research projects. An excerpt from his letter of October 9, 1939, speaks to the point:

"So I venture to hope that our notes may show increasingly other observations than those which relate intimately to our special problems of personal interests. Naturally I realize that we all consider ourselves busy and can readily plead that as an excuse, but actually it is no excuse because it is wholly a question of importance and precedent. If we are going to bother to keep life-history records of all individuals and to further the solution of varied biological problems, we cannot afford to neglect our daily opportunities nor can we place the burden upon our colleagues. It is responsibility which we should share equally. Naturally I am more acutely aware of credits because each month I have the opportunity to read consecutively all of the records North and South. I am minded to try some procedures for increasing interest in this matter and improving our results".

"All this is not in criticism of the Station's reports, because in general they tend to be more satisfactory as matter of miscellaneous records than the New Haven notes. But in both instances we can do infinitely better, so I am thinking rather of improvement than of shortcomings".

I was easily convinced of the potential value of note taking and, because I enjoy descriptive writing, could look upon it as much more than an expected laboratory chore. From my correspondence one gets a clear impression that when these

reports came to Doctor Yerkes' desk he dropped everything
else to read them. If they were late in arriving we heard
about it. If an individual presented a meager amount of
material he howed about it. The remainder usually carried the
charitable assumption that the individual must have been ill.

Although there was some pressure upon us to "turn out
copy" I know of no instances of padding nor recording of
fictitious events. I continue to find, in reviewing my own
daily notes from the 1930 to 1939 period, many cues for further research. In addition, these notes should have much of
interest to social psychologists, parents, pediatricians and
others. I propose here to give a few samples from these notes
which deal largely with social relationships and personality
characteristics.

It is tempting to start with the babies, their differences in feeding patterns and behavioral development, but we
might get bogged down in infancy and have time for nought
else.

Likewise, it would be easy to overspend our time with
the children and adolescents on Prospect Street, although
these six immigrants, fresh from Africa, eating and sleeping
within the same modest quarters, provided a rare opportunity
for observation. It was a dynamic social situation with gang
warfare, sudden and puzzling shifts of alliances, impressive
expressions of group cohesion and mutual affection. I tried
my best to understand what was going on and even though I
seldom found out I felt it my responsibility to play the role
of policeman.

My efforts at peacemaking were usually met by a chorus
of hoots. I rarely solved any problems. Years later I often
recalled these experiences when my own children, aged 7 and
10, engaged in disturbing brawls. Oddly enough I was still
using the same technique that had never worked well with the
chimpanzees. On one particularly noisy occasion I intervened with the intent of imposing some just solution. This
time, instead of being greeted with hoots, the ten year old
said, perhaps with a stamp of her foot for emphasis, "Daddy,
you don't know a thing about this situation!" How true, and
I quickly realized that I never would discover the facts from
either participant. I can imagine that those hoots back in
1930 were giving me the same message. Nevertheless, my child
rearing practices underwent an abrupt and permanent change
following this experience.

Most of the diary accounts of these social conflicts
are too long and involved to include here. One of the shorter
entries must suffice.

January 22, 1931

Attracted about middle of the afternoon by prolonged
screaming from Moos. Peeped through the door and found Song
attacking him. Boy was in the mix-up, taking the part of Moos
by slapping and pulling at Song. Upon my entering the room
Song scampered to the bar near the ceiling. HWN followed on
the bar until we left, then came down immediately and began
making a nest in the straw on the floor. Boy and Moos had
climbed across the window to the bar in the meantime. Boy
became stranded on the bar. He had hung on until he was unable to swing back to the window. Began a low whimpering
call, whereupon Song got up from his nest-making, went over
and reached an arm up the bar. Boy threw his arms about Song
who dropped to the floor. On the part of Song, this was a
very quickly executed and effective rescue.

Later, from the Orange Park Laboratory, I select the
following entries:

February 11, 1935

Dita - Exchanging objects for food. At about 8:00 PM
Nana was discovered with one of the aluminum binding wires
which she evidently had pulled from her cage. After obtaining slices of orange I spent several minutes trying to induce
her to trade the wire for the fruit. She refused and became
more careful about thrusting the wire within our reach. At
3:30 she pushed the wire through the partition and Dita se-
cured it. Instantly, she darted to the front of the cage and
gave it to me, even before I had made any move to trade fruit
for it. Having given up the wire she held her hand for reward which she was given. Previous mention of such behavior
on the part of Abreu animals is to be found in notes.
(General file, Aug. 2, 1931 and individual files). Today's
observations seemed to be especially significant because of
long retention of this response to food exchange response.
Desiring to repeat the observation, the wire was returned to
Nana who again refused to give it up. Several times where
Dita was shown food she looked toward Nana and the wire. She
made a few unsuccessful attempts to take it from Nana. Finally, after wire had accidentally been lost by Nana and re-
covered by us, Doctor McGraw took it to far corner of cage.
Dita was facing me as I showed her a piece of orange. Instead
of coming directly to me (the usual response to food) she ran
to the opposite corner, got the wire and presented it to me.
This was repeated three times. Doctor Myrtle McGraw, Doctor
Spence and Mr. Lawrence Frank also observed the above behavior.

July 20, 1935
    Wendy - Temperament. Wendy was discovered with the hose
which she had drawn through the wire. She became furious when
I pulled it away from her. I went about other tasks and when
I passed her cage again, 10 minutes later, I had forgotten the
incident. She was sitting on the ledge with a mouthful of
water which she drenched me with. The behavior is worth noting
because Wendy rarely spits water at attendants - although she
frequently does at animals in adjacent cages.

September 18, 1936
    Mona - Exchange of infant for food. Since Mona frequently
presents various parts of her infant through the wire in order
to obtain food I thought it would be interesting to see what
she would do when she had the opportunity to present the entire
baby. Would she actually release the baby for a _large_ food reward? Working through a six-inch opening of the door she
seemingly made genuine attempts to hand the baby through but
always, as it lost its hold, it would begin to cry and Mona
would draw it back. I suspect that she might not have allowed
it to be taken even though it had been quiet, but she presented
a good picture of cooperation nevertheless.

    It was a general policy in the laboratories to discour-
age visitors of the sightseeing variety, but there were a few
who gained entrance by special permission. Those having little
acquaintance with the anthropoids asked many questions, the
strangest of which was: "How do you tell them apart?"
    Unless chimpanzees have ceased to show individual diffe-
rences, I'm sure the range of personality characteristics in
the present colony is as large and as varied as in the early
groups in New Haven and Orange Park. Nevertheless we had
some colorful characters. Without consulting with previous
colleagues who knew her well, my choice for the most interesting personality would be Fifi. I use the word "interesting"
deliberately rather than other descriptive terms. We had some
psychopathic individuals but I would not include Fifi in that
group. She was interesting.
    In my letter of November 25, 1935, to Doctor Yerkes, I
wrote of Fifi: "It is my firm belief that she is the most outstanding liability in the colony and the sooner she arrives in
the hands of an able pathologist the better. It would not be
difficult to assemble evidence to prove that we can't afford
to keep her solely as breeding female. Of course we shall
continue to care for her as best we can."
    An immediate reply advised us that if we had ruled out
intestinal parasites as a major disturbance we should try
psychotherapy. "She is a highly social individual, long

accustomed to much sympathetic attention. Why not place her
in a congenial social environment and suggest to everyone about
the station that she be given special attention? ... I am not
inclined to share your present pessimism ... Tact as well as
sympathetic social insight I should suspect to be of first rate
importance. Good Luck."

Accordingly we placed Fifi in a conspicuous corner cage,
instructing all laboratory personnel to say "Hello" to her as
they passed by. Our concerted attention apparently was not
enough to satisfy Fifi. She began bumping the loose hanging
steel door with her shoulder, creating a din quite disturbing
to persons analyzing data and writing papers in adjacent laboratory offices.

To cut off this racket I drove small wooden wedges between the door and its frame. It was only a matter of minutes
until Fifi had removed the wedges and the clanging started
anew. My second attempt to drive in firmer wedges was greeted
with a first class temper tantrum.

Although a few others showed active interest in clear
mirror reflections, Fifi was the only one who used pools of
water on the floor after cage washing. She would draw her
fingers through the pool, peering closely at the distorted surface image, excitedly smacking her lips and stamping her right
foot as she watched. With no water in the cage she has been
observed to make her own puddle and proceed to amuse herself.

The diary notes contain the following observations:
April 20, 1937
    Fifi - Behavior with hair-clippers. A pair of hair-
clippers was just outside of living-room cage as the door was
opened and Fifi entered. She went straight for them and reached
them before I did. I spent some time trying to trade her food
for the clippers. Ordinarily she is a good trader, but in this
case she ignored me entirely. For an hour she had a grand time
taking the clippers apart, examining, and arranging the pieces
on the floor. When lunch-time came, and she had apparently
tired of her plaything, she promptly returned all the parts for
food, even going back to look for other small pieces when
ordered to do so.

A final diary note on Fifi is entitled "Theft of Soda's
Infant". It was the first and only case of kidnapping we had
at the station. Instead of presenting the excessively long
diary account of the incident a paragraph from my letter of
October 4, 1937, to Doctor Yerkes gives an adequate summary:

"Perhaps among the predictions you might make concerning
the range of social behavior in chimpanzees, kidnapping would
not be one of them." Fifi gave us a very busy morning yesterday after stealing Soda's baby and refusing to return it to
her. It was a rather laughable situation from our point of

view, but at the same time perhaps from Soda's point of view
no joke. We soon separated the two adults, and after several
attempts to induce Fifi to part with Coma we separated them
in the separation box. This incident has caused us to keep
Fifi and Soda in separate cages. They have always been quite
friendly and seemed to get along unusually well together and
might do so again for an indefinite time, although a future
occurrence of this kind might not turn out quite so favorably
as it did yesterday."

The reply from Doctor Yerkes was that he was impatient
to get the full details. These were later supplied in a
special report.

Those of you who knew Doctor Yerkes need not be reminded
that his involvement with the laboratories, North and South,
was total. He did encourage and participate in fun and games
on a scheduled basis. Frequent staff parties were given in
the Yerkes home at 4 Saint Ronan Terrace. At Orange Park the
main outdoor game was badminton in which most of us participated.

It was my impression that for Doctor Yerkes games and
other recreation activities were lively interludes in which
he could participate while thinking about the design of a new
addition to the laboratories. By his own account in Murchin-
son's History of Psychology in Autobiography he denied a reputation he had among his friends and family that he was a
hard worker. It is true that he had a rather short work day,
but hard work should not be defined in terms of hours spent
on the job. His time schedule for work and play, except for
unusual circumstances, was a tight one. There is no satisfactory way of measuring short-term output of the type of work
Doctor Yerkes did. If we look at the long term we must believe
that he made the most of those short work periods - otherwise
we wouldn't be here today.

These reminiscences must be concluded with some remarks
on an apparently guiding principle of Doctor Yerkes which contributed substantially to staff morale. His interest and concern about the general welfare and progress at the Orange Park
Laboratory was increasing during the late 1930's. During his
usual period of residence in New Haven we carried out a volu-
minous correspondence about happenings. This was in addition
to the monthly diary reports. Only very rarely did he communicate by phone as one might expect now in a similar situation.

In most of his letters his concern was clearly evident
by his repeated and extensive admonitions. We responded to
these as best we could, even though we felt that many of them
was unnecessary. However, when tragedy struck as it sometimes
did, especially in the winters of 1937 and 1938, there was no
accusations of neglect of duty or intimations of blame.

In all of his letters during those trying periods there is not to be found a single note of criticism or blame, even after such tragic events as the loss of Peter from thermic fever. (He was a prize infant, as the first born of Cuba, the first chimpanzee in the world of known birthdate).

What I have just said is best understood in his own words. His letter of October 24, 1936 contains the following:

"I appreciate that the station staff has had a difficult and discouraging time for some weeks. It is small comfort to know that such experiences, or worse, are common in life. To hold steady and avoid worry and sense of affliction at such times may be almost impossible - I seldom say much about our heroism or patience because in the last few decades I have come to take them for granted in any one whostays long with this group. It is the nature of our job to demand more devotion and self-sacrifice and discouragement than we feel equal to... It is not that I am unaware of the nature of the situation and its demands, or that I do not take account of the quality of the services of the individual. Ours is essentially public service because we are working for mankind, not for one another, or the university. I habitually think of the group as motivated by love of service."

Again, from his letter of December 1, 1937, following an epidemic and some casualties in the colony, there is this excerpt:

"Cheers certainly are not in order, but cheering sympathy doubtless is. You must have had a well nigh impossible time for the last two or three weeks, and it does seem as though worries, perplexities and disappointments in our job were endless and also tend to bunch."

Following both of these rough periods of 1936 and 1937 at Orange Park one might expect that he would make the common response of telling us what we should have done. He never did. And for this we are deeply grateful.

REMINISCENCES OF EARLY DAYS IN NEW HAVEN
AND ORANGE PARK

S. D. S. Spragg

University of Rochester
Rochester, New York

Mistakes sometimes lead to favorable outcomes and my getting to the Yale Laboratories of Comparative Psychobiology illustrates this. In the winter of 1932-33 I was a young graduate student in psychology at the University of Washington in Seattle. I was doing some research in animal behavior and wanted to go to an eastern university to complete my graduate training, so I wrote letters of application to several institutions. In writing to Yale I had assumed that Dr. Robert M. Yerkes was chairman of the psychology department, since he was the psychologist there I had heard most about. Thus I wrote to him.

I received a prompt and courteous reply, pointing out that he was not chairman of the psychology department, but that if I were interested he would be pleased to consider my application for a graduate assistantship in his Laboratory of Comparative Psychobiology. He then described the laboratory and its activities. I responded eagerly and was delighted to be accepted as research assistant in his laboratory and as a graduate student in the psychology department.

I arrived in New Haven one afternoon about the first of September, 1933 and met Dr. Yerkes and his two staff colleagues in the laboratory, Dr. Henry W. Nissen and Dr. Carlyle F. Jacobsen. They and others in the lab were about to depart for the annual APA meetings and I was told that I could be of service by assisting Dr. Yerkes' son David, a handsome young man of about twenty, in feeding the animals their morning and evening meals while the rest of the lab staff was away at the meetings.

That first afternoon I put on a borrowed lab coat and went up to the animal living quarters on the top floor of the Sterling Hall of Medicine to observe the feeding procedures, which took place in a smallish room containing a solid wooden

table with small seats attached. The door to the cage was
opened, four dark-faced young animals came out and went at
once to the table, climbed on the seats and started to drink
a milk and porridge mixture from metal cups. Suddenly one of
them (I believe it was Kambi) got down, walked quietly over to
me, bit my hand, and as quietly walked back and resumed her
meal. That was my introduction to chimpanzees.

The problem which I proposed for my dissertation research
had to do with anticipation as a factor in maze errors, a development of some research I had done with rats at the University of Washington. Dr. Yerkes approved the proposal and told
me I would be working on it under Dr. Nissen's direct super-
vision. I believe that mine was the first doctoral thesis
problem in the lab that had not been largely assigned. Of the
three that preceded mine, Kenneth Spence's had been on visual
acuity, Jim Elder's on auditory acuity, and Milton Forster's
on simple reaction time in young chimpanzees.

In my dissertation research I used serial learning tasks,
training five young chimpanzees to work at channel-type stylus
mazes, both spatial and temporal. During part of the study
subjects worked at the mazes with a blindfold over the eyes,
which required a good deal of preliminary training.

In the summer of 1934, one year before I finished my
degree, I spent a month at the Orange Park laboratories at
Dr. Yerkes' invitation. During that period I roomed and
boarded in the Yerkes home (as did also Helen Morford, Dr.
Yerkes' secretarial assistant). For a month I had virtually
all my meals with Dr. and Mrs. Yerkes and also drove to and
from the lab each day - morning, noon and late afternoon --
with Dr. Yerkes and Helen. On Saturdays after lunch the four
of us would usually drive into Jacksonville for various er-
rands followed by dinner at a restaurant in town.

Thus this was a month of frequent and close contacts with
Dr. Yerkes. I found him to be uniformly courteous and willing
to talk about a wide range of topics, but typically rather formal in his conversation. During the four years I was with him
he always addressed me by my last name, usually preceded by
"Mr.", then "Dr.", but in informal settings it often became
simply "Spragg". This is in marked contrast with recent cus-
toms, in which most of us call our graduate students by their
first names from the time of their appearance, and many of them
within a few weeks are calling us by ours. I am confident
that Dr. Yerkes would not have been comfortable with this. My
impression from this period was that he was essentially quite
a shy person in most situations and that his formality was, at
least, in part, due to this.

After receiving my Ph.D. degree (June, 1935) I spent two
years as a post-doctoral research assistant in the Yale

Laboratories of Primate Biology at Orange Park. Dr. Yerkes
had invited me to carry out a major study of morphinism in
the chimpanzee, with special reference to indications of
addiction, e.g., evidences of desire for the hypodermic injection and its results.

Before beginning the hypodermic administration of morphine, I trained four young chimpanzees to submit without
restraint to a needle prick, then to hypodermic injections of
neutral saline solution. As an aside, I might mention that
Dr. John Bruhn and I similarly trained these animals to submit
to a needle prick on the finger to obtain blood samples for
blood counts. The morphine study lasted approximately two
years, usually one dose schedules of seven days a week, with
occasional periods of withdrawal to assess the development of
addiction. We were able to demonstrate the development of
strong and unambiguous addiction, with many similarities to
human morphine addiction.

Dr. Yerkes followed the study closely and as a matter of
fact personally took over the injections and experimental observations for a one month period in the summer of 1936 while
I went north to be married. He and Mrs. Yerkes very generously
offered us the use of their farmhouse in Franklin, New Hampshire
for part of our wedding trip, and we spent some pleasant days
there getting acquainted with the fields where Pan and Wendy
and others had romped.

During my stay in Orange Park Dr. Yerkes and I collaborated on a study which was published by invitation in 1937 in
the Journal de Psychologie with the title "La Mésure du com-
portement adapté chez les chimpanzés". The study was done
with the intent to standardize three learning tasts -- a box
and pole task, a "detour" task, and a task requiring box using
(and stacking) to obtain a suspended lure -- with the hope
that they might prove to have some general normative value in
assessing the learning ability of individual animals. However,
it did not become a chimpanzee "Binet test" by any means, and
I believe that one reason the paper remained rather obscure
was because it was published in a French journal.

Also, at Dr. Yerkes' request, I gathered finger and hand
prints of most of the animals then in the Orange Park colony,
adults as well as juveniles. The results were published, with
Dr. Harold Cummins of Tulane University as senior author, in
1938 in Human Biology with the title "Dermatoglyphics in the
Chimpanzee". In getting prints from the adult animals, I
trained them to thrust a hand and fore-arm through a special
opening in the cage door, then permit me to press a finger or
the entire hand against a suitably mounted sheet of chemically
sensitized paper. I was keenly aware that an adult animal,
especially one of the large males, could easily have taken my

hand and arm back through the opening if he had wished to do
so, but fortunately no such untoward incident occurred.

During this period I developed a recording form and
procedures and made systematic descriptions of skin color and
hair color for all the animals in the colony. This material
was made available to Dr. Harold Coolidge of the Harvard University Museum of Comparative Zoology for his analysis in
connection with his comparative studies.

My two years at Orange Park were busy, productive, and
enjoyable ones. The young investigators at the laboratory were
a closely knit group, socially as well as professionally. There
were about a half-dozen of us in a small hamlet of some 200
souls, mostly poor blacks plus the residents of a retirement
home. We learned a lot during that period -- from the chimpanzees and from each other, and also from Dr. Yerkes' experience and wisdom. I am grateful for the experiences I had
there, and for the opportunity to know and work with Robert M.
Yerkes, that great pioneer and developer of the study of
primate behavior.

MEMORIES OF MY FRIEND ROBERT M. YERKES
AND HIS "PRINCE CHIM"

Harold J. Coolidge

38 Standley Street
Beverly, Massachusetts

It was my good fortune to have my own interest in the
Anthropoid Apes, and particularly gorillas, in a large measure
sparked by a great biologist, my distinguished friend, Robert
M. Yerkes, for many years Professor of Psychology at Yale University.

I not only visited him and his helpful wife Ada in New
Haven on several occasions to discuss primate research, but
also used to stop at their farm in Franklin, New Hampshire on
my way to our own summer island camp on Squam Lake. For me he
was a true father figure. He was always warm and friendly and
deeply interested in my various activities. It was at his
hill farm that I had the privilege of meeting their charming
small eared, black faced, Prince Chim, whom they acquired in
August, 1923, and who was rightfully described in Yerkes tribute to him as "extremely intelligent". He was notable for
his bold aggressive manner, his constant alertness. He was
even tempered and always ready for a romp. In fact, he was a
great showoff, and I recall his welcoming me with a hearty
hand-shake. It was very sad that this promising little four
year old chimp fell ill, and met his untimely death while
visiting Madam Abreu's famous Quinta Palatino colony in Havana
in July, 1924. He deserved the Yerkes tribute title of an
"intellectual genius" as described in the author's book,
"Almost Human".

Little did I realize that in my own 1928 survey of chimpanzee skeletons and skins in European museums I should discover in the Congo Museum in Tervueren, Belgium four skulls
and skins of a new pedomorphic chimpanzee from south of the
Congo River described shortly afterwards in October, 1928, by
Ernst Schwartz as "Pan satyrus paniscus", and subsequently,
on the basis of my own research, raised to the status of a

43

full species in my 1933 report on "Pan paniscus", published in
the <u>American Journal of Physical Anthropology</u>. In this I note
that the reported locality of origin of Prince Chim near Lubu-
to placed him on the East bank of the Lualaba River within the
Paniscus range, that a study of his skull indicates a possible
age of 5 or 6 years, which may help to account for his greater intelligence as compared to a younger animal. In general
skull characters and brain capacity he resembles most closely
"Pan paniscus". This leads me to conclude that he was probably the first specimen of this new species in any American
museum. How wonderful that Professor Yerkes was the first to
identify so closely in his accounts of Prince Chim many of the
behavioral characteristics of this remarkable new species,
that have since been increasingly confirmed by further studies
as additional specimens have become available to psychobio-
logists and ethologists.

My discussion with Professor Yerkes in my Junior year at
Harvard kindled my interest in gorillas, and when I spent my
Senior year crossing Africa with the Harvard Medical Expedition; led by Dr. Richard P. Strong, one of my assignments was
to collect a large specimen of the mountain gorilla for the
Harvard MCZ. This led me into a study of all accounts of
gorilla hunting in the literature and talks with Carl Akeley,
Jim Chapin, etc. Carl sadly died of dysentery shortly after
we arrived in the Kivu area, before we could meet him there,
and was buried in the gorilla habitat in the Virunga Volcanoes
of the Eastern Congo.

I was fortunate to be able to bring back a fine male
gorilla specimen for Harvard's MCZ, with preserved organs, and
very complete measurements. He was shot in the Katana Mountain area west of Lake Kivu. My field exposure led me later
the following year to a study of gorilla collections in museums in Europe and the U. S. which resulted in my monograph on
the revision of the genus gorillas published by the MCZ in
1929 in which I was able to reduce described races from 36 to
2 which, happily, have been largely accepted by scientists
ever since. This research was inspired by my Harvard mentor,
Professor Glover M. Allen, as well as by Professor Robert M.
Yerkes. The latter indicated the need for organizing a similar study of the classification of chimpanzees, which I undertook in 1931 by studying chimp collections in the leading
museums of Europe and this country, as well as inspecting living anthropoids in the larger zoos in addition to the Yerkes
chimps at Yale, and at the new field station in Orange Park,
Florida.

One of my memories of several days spent there, where
Dr. Nissen was working at the time, was Professor Yerkes
telling me of his problems in establishing a good personal
relationship with human neighbors. This was made difficult by

the fact that the general public had to be kept out, as this
was not a zoo, but a place where serious research required
tranquility for the animals in the colony. He said that a
story was going around that the Orange Park Station was a secret way of influencing the local elections, because a large
number of controlled votes could surely emerge from this locked installation established by Northern Yankees!

I regret to say at this point that while I gathered an
immense amount of data on chimpanzee material in museum
collections, and I also published the first note on two clear
cases of supernumerary mammae, I have never completed my monograph because of the press of other duties; and in part because Dr. Ernst Schwartz brought out a paper on his own ideas
of a revision of chimpanzee classification, which largely
agreed with my own findings, but was based on far less material.

My own efforts with Professor Yerkes encouragement to
get more primate studies underway at Harvard resulted in
organizing a course on "The Evolution of Animal Sociology"
under Professor William Morton Wheeler. The span from amoeba
to man was covered by eight different lecturers, and Professor Hootah and I dealt with the higher primates, including
early man. The best required reading for our selection included Robert and Ada's splendid book on "The Great Apes."

The momentum developed through Professor Yerkes stimulation of my primate studies led to the organizing in 1937 of
the Asiatic Primate Expedition primarily to study the gibbon,
the least known of the anthropoid apes, as well as the
orangutan through field studies and collections in Siam,
Borneo and Sumatra. This proved highly successful as a result
of hard work by Adolph Schultz of Johns Hopkins, Ray Carpenter
of Columbia, Sherwood Washburn, Gus Griswold, and myself
(prior to illness) from Harvard. The papers that resulted
from the field and laboratory studies of our collections greatly increased our knowledge of the gibbon, and incidentally
a new tree shrew from Mount Kinabalu in North Borneo.

Upon our return from the field a further effort was made
to establish a Primate Laboratory at Harvard which Professor
Yerkes had always encouraged, and plans were drawn up by
Schultz, Washburn and Carpenter. Funding problems prevented
its accomplishment, and further efforts were frustrated by the
outbreak of World War II.

It is, however, gratifying to report that working through the International Union for the Conservation of Nature
which was founded in 1948, positive progress was made in
establishing and strengthening natural habitat areas for the
protection of wild primates. Outstanding among these was
extending the "Albert, Kivu or Virunga National Park" in the
volcanoes of the eastern Congo where Carl Akeley was buried.

Other primate habitats in African Parks and Reserves have also been established, and the Survival Service Commission on the IUCN maintains a watching brief to identify situations on a world-wide basis, such as that of the Golden Lion Marmoset, that require immediate governmental action for habitat protection.

     There is an old Chinese proverb that in paying a person a compliment you say "May your shadow never grow less!" In Professor Yerkes case it would please him to know that the shadow of the great name he has cast in the field of primate studies has continued to grow. I am sure that all of us at this Memorial Conference owe him an immeasurable debt of gratitude for initiating the rising tide of primate knowledge to which we each try to contribute in our own small way!

REFERENCES

Almost Human - The Century Co., 1925, pg. 278, R. M. Yerkes.
The Great Apes - Yale University Press, 1929, pg. 652,
     R. M. Yerkes and Ada W. Yerkes.
Chimpanzees - Yale University Press, 1943, pg. 321, R. M.
     Yerkes.
Le chimpanze de la rive gauche du Congo. Bull. cercle Zool.
     Congolais, v.70, 1928 - Ernst Schwartz.
Das Vorkommen des Schimpansen auf den linken Kongo-Ufer.
     Rev. Zool. Bot. Afr., XVI, 4, 425, 1929 - Ernst
     Schwartz.
A Revision of the Genus Gorilla. Memoirs of the MCZ at
     Harvard College, v.50, No. 4, 1929 - pg. 295-381,
     H. J. Coolidge.
Symmetrical supernumerary mammae in a chimpanzee. J. Mamm.
     v. 14, No. 1, Feb. 1933 - pg. 66-67. H. J. Coolidge.
Pan Paniscus. Pygmy chimpanzee from South of the Congo River.
     Am. J. Phys. Anthrop., vo. 18, No. 1, July-Sept., 1933.
     pg. 59, H. J. Coolidge.
Zoological Results of the George Vanderbilt African Expedition
     of 1934. Part IV. Notes on four gorillas from the Sanga
     River region. Acad. Natural Sci of Philadelphia.
     Vol. 88, 1936. Published, Oct. 23, 1936, pg. 479-501.
     H. J. Coolidge.
A new tree shrew of the genus Tana from Mount Kinabalu, North
     Borneo. Proc. New England Zool. Club, Vol. 17, pg. 45-47,
     (May 6, 1938). H. J. Coolidge.
Robert Mearns Yerkes. (1876-1956). Yearbook Amer. Philosophi-
     cal Soc., 1956. pg. 133-140. Edwin C. Boring.
Robert Mearns Yerkes. (1876-1956), J. Psychological Review
     vol.64, No. 1, 1957. pg. 1-7, Leonard Carmichael.

Pygmy chimpanzees. Oryx, Feb. 1977 (in manuscript), Alison and
    Noel Badrian (6 month field study of Bonobos).
Preliminary Information on the Pygmy Chimpanzee, "Pan paniscus"
    of the Congo Basin. Primates 13(4), pg. 415-425, 1972,
    T. Nishida.
Mountain Gorillas and Bonobos. Oryx, July 1976, pg. 372-382.
    John MacKinnon.

DR. ROBERT YERKES AND SEX RESEARCH

Vincent Nowlis

University of Rochester
Department of Psychology
Rochester, New York

Dr. Yerkes often said that a major reason for placing the Yale Laboratories of Primate Biology in Orange Park was to establish a facility for a successful breeding program and co-incidentally, a program for study of chimpanzee sexual, reproductive, conjugal, parental and other social behavior in a setting somewhat more like the natural habitat than could be set up at that time in a colder climate. These programs demanded diverse daily attention and work from those of us working with him in Orange Park and later with Dr. Henry Nissen, Among the full-time Orange Park old timers present today, almost every one published studies that were part of what turned out to be successful programs.

Shortly after leaving Orange Park in 1942 to try to learn to teach, as well as do research, at the University of Connecticut, I received a letter from Professor Alfred C. Kinsey, of whom I had never heard, saying that Dr. Yerkes had recommended me as a junior colleague in a research project on human sex behavior. Manifest qualifications were my training in experimental and observational study of behavior and a continuing interest in primate socio-sexual interaction. After an encouraging message from Dr. Yerkes, I had interviews, including the inevitable sex history, with Kinsey, and with Pomeroy and Martin, all three of whom shared a most unexpected, refreshing and engaging enthusiasm for sex research and a broad and evenhanded perspective on sex and research. Despite such negative characteristics as my then poor handwriting (Yerkes never would have made it!), relative youth and, in the Kinsey perspective, inexperience, I was invited to join the tiny staff for a minimum period of two years, which I did (Pomeroy, 1972, pp. 158-162). I introduce my brief comments in this way because this position put me, as friend and former student of Yerkes and as colleague and post-doctoral student of Kinsey, in the midst of some of the

are objective to an astonishing degree when one realizes the
complexity of the problem he is undertaking. His very method
of recording not only assists in keeping the information con-
fidential, but also yields observed material that is capable
of quantitative analysis", (Pomeroy, 1972). The Committee's
subsequent and continuing approval of this project (and thereby
the implied approval of the National Research Council and the
Rockefeller Foundation) made for greater legitimacy of the project, giving Kinsey access to significantly more and more populations in his quest for the Grail; that is, for the coveted
100,000 informants who, 50 or more deep, would fill the myriad
break-down cells for age x sex x ethnicity x marital status x
religion x devoutness x residence x education x socio-economic
status x region, etc. Financially, the Committee increased the
$1600 to $7500 (1942-43) and thereafter allocated about 50% of
the total budget to the Kinsey project.

From this point on there was frequent correspondence.
Kinsey often asked for advice on specific tactics and Yerkes
would reply constructively, adding words of advice on overall
strategy, which Kinsey then courteously acknowledged and often
rebutted. A basic issue involved Kinsey's adamant belief that
he must interview and include in the series all available informants and exclude none because of an unusual sexual history.
By contrast, as late as 1933, Yerkes had approved in the
Committee's annual report the statement that "sex patterns ...
can be better studied among ourselves", and in early exchanges
with Kinsey tried to persuade him "not at this time to get
into studying 'perversions' but to stick to a modal group or
to a cooperative, representative group like college professors.
But Kinsey felt that "colleague professors represent one of the
widest departures from anything that is typical of the mass of
the population" (Pomeroy, 1972) and that the concept of a
modal group in sex research was meaningless. It was not long
however, before Dr. Yerkes acknowledged the "fairness and
wisdom" of Kinsey's defense of the scope of his undertaking
and added: "The situation is so complex that one must be pragmatic always, and also so far as practicable, logical!"
(Pomeroy, 1972, p. 168).

Yerkes' hard-gained wisdom provided sounder and more ger-
mane advice on several broader issues. He wrote:
"It is my conviction that your value to the project
to which you have dedicated your life will be measured
in future decades rather by your breadth of view,
prophetic insights and wisdom, than by the quantity of
your work and its representation in factual reports.
[Moreover] .... develop a staff organization which
would assure optimal research progress while permitting the overall direction to take in good conscience
profit from frequent periods of vacation ... and

elsewhere in Bloomington!"
(Pomeroy, 1972, pp. 168-169).

Alas, much of the best of Kinsey -- most of the insights, the wisdoms, the breadth of view -- was never published. And no vacations led to a too early death at age 62.

I turn now to a pleasant visit to the primate section of the St. Louis Zoo at the time of AAAS meetings in that city in, as I recall, 1946. Both Yerkes and Kinsey were attending and I easily persuaded them to make the visit together (Nowlis, 1965). Despite their lengthy discussions about primate sexual behavior, they had not yet had any occasion to observe together and to discuss those observations. I knew by that time that both men were excellent observers of animal behavior but I also had a feeling that any one item of social behavior might produce divergent perceptions in them, a hunch that had humorous support during our short visit.

As investigator of human sex behavior, Kinsey sorely felt that he had an obviously crippling handicap in the fact that he had to depend on verbal report for his basic data. As psychobiologist, he yearned, like Euclid, to see beauty bare. Thus the tremendous drive to get photographic and other graphic, revealing and long-to-be-studied records -- erotica, graffiti, pictures, cinema, drawings, paintings, sculpture, books, records, artifacts, of all kinds. He witnessed sex shows. He had colleagues in gynecology and urology make observations, so to speak, in vivo. Like a Ray Birdwhistell, he strove to remove the perceptual filters with which his culture had additionally handicapped him in order to identify the whole range of behaviors in sexual repertoires, from the tiniest to grossest item, from the ephemeral to the enduring. He later added to the staff a professional photographer largely to get motion picture records of animal as well as human sexual behavior and his first recording session was at these laboratories in Orange Park.

By contrast (as Yerkes would say), Yerkes seemed almost to strive to organize as much as he saw in chimpanzees social interaction into perceptual categories that were culturally congenial and significant to him. His writings on such interaction abound with helpful terms like right, privilege, custom, status, altruism, leadership, consort. I think he may have been very pleased with a social psychology of the chimpanzee in which the major concepts of cultural anthropology were operationalized in terms of observable chimpanzee behavior.

As we three slowly proceeded from one primate group to the next, stopping for a period of discussion wherever there was active or interesting social interaction, I both watched and listened. Both men were pleased with the healthy appearance of the animals and with their lack of boredom with each other. (Marlin Perkins was, I believe, still the director but was not at the zoo on the day of this unplanned visit). Kinsey's delight with these "not too poor" relations of man was shown in many ways. The enthusiasm was apparently based, in part, on his seeing and feeling sexual undercurrents and nuances (to him of obvious potential importance for any science of sex) in much of the interaction we watched: glances, movements of the eye, mouth, lips, tongue and other bodily parts, vocalizations, gestures, casual contacts, prolonged grooming, bodily inspections (visual, olfactory, gustatory), social initiations, rejections, rhythmic dances, conflict, cooperation, sharing and more. Here the healthy sexuality, manifest and implied, seemed to him so copious and diverse and, often, fleeting and subtle, that Kinsey could only hope that someday our observational categories and methods could provide accurate and significant sets of knowledge about these behaviors and experiences.

Yerkes also enjoyed the afternoon. He gracefully played the role of experienced observer and old friend of primates, one who points to that which he has seen before, as now. He was fully attentive to both Kinsey's stated observations and to what, for Yerkes, were Kinsey's obiter dicta. That wonderful facial blush of his more than once suggested he had not seen what Kinsey had. Accordingly, the always tactful Kinsey adjusted his discourse and an instructive and comfortable experience was had by all, each in his own way.

Finally, here is a passage in which Yerkes, with George Corner, describes an historical relationship between Freud and Kinsey, the two best sex researchers of our time, a relationship in which he was an active agent:

"In the United States, the twentieth century has been a period of exceedingly rapid and revolutionary change in sex attitudes and practices ... These changes made feasible the work of the Committee, and prepared the way for the Kinsey report. Only a few decades earlier, Henry Havelock Ellis (1859-1939), eminent student of sex in England, suffered severe censure and legal restrictions of the publication of his scholarly findings. Only a little later, Sigmund Freud's (1856-1939) account of the role of sexuality in the ethiology of the neuroses was rejected by his medical colleagues. But the Austrian psychiatrist persisted with his inquiry, and eventually proposed theories and formulated hypotheses which have fundamentally changed our conception of the role of sexuality in our mental and social lives.

Comparison of Freud and Kinsey are not implied, for the two men differed greatly in temperament, professional training and experience, and in their objectives; but what should be noted is the fact that Freud, on the basic of his clinical experience, proposed theories

which laid the foundation for a task he was not fitted
by nature or training to carry on. This is the great
task of fact-finding through careful, patient, long
continued, objective research which Alfred Kinsey,
the laboratory and field-trained biologist, is now
engaged in doing. From the Kinsey project, sufficiently extended, should come basic knowledge against
which theory may be checked, modified and supplemented." (Kinsey, Pomeroy, Martin and Gebhard, 1953,
pp. 7-8).

It is important to note that in 1925 the Committee sent
Earl Zinn, one of its organizers, to visit European scientists
and physicians, including Freud, but Zinn's reports on the
trip were never published and were apparently preserved only
in the "heads of the committeemen". Aberle and Corner (1953,
p. 6) say: "Freud was pleased to learn that a group of outstanding American scientists was interested in the investigation of sex and named one of his pupils in the United States
as a possible collaborator with the Committee. The suggestion
was not followed up, apparently because psychoanalytic concepts
were not considered to come within the scope of objective research of the sort represented by the Committee."

By 1954, we find that Kinsey, in a reference to the
Yerkes-Corner passage quoted above, believed as follows: "It
would be interesting to note that the relations between our
work and Freud's depends upon specific recommendations that
Freud made in the middle of the 1920's to the American group
that was first organizing the National Research Council's
Committee for Research in Problems of Sex. He very definitely
saw the need for gathering the sort of factual record which
we have accumulated. Dr. Yerkes, who was chairman of the Committee when Freud first made these suggestions, summarizes the
relation between Freud and our work in the Foreward to the
present volume." (Pomeroy, 1972, p. 356).

Freud asking for "fact-finding through careful, patient,
long-continued, objective research"? Yes, but only in other,
more general contexts, as far as we now know. Yerkes asking
for such? Yes, and in this specific context. What we clearly
see is Yerkes as a major and timely and deliberate and modest
and active mediator between two of the very significant scientific events of his life, the work on sexuality of Freud and
that of Kinsey.

In conclusion, I sadly note that the Committee in the
early 1960's carefully brought about its own dissolution for
reasons described by Frank Beach in his Sex and Behavior. The
major reason was financial; private funds dwindled while the
grant programs of the NSF and NIMH grew. "In the face of this
welcome competition, the committee ... came to the inevitable
conclusion that it no longer had any unique function to serve

as far as direct support of research was concerned, and no new grants were made after 1961." The negative consequences of this decision and of its precipitating circumstances are greater than anyone might have anticipated. In brief, as of 1976, we find the domain of human sex research on a plateau only slightly higher than that at last reached in the late 1940's by the Committee after many years of wandering efforts at very low levels. We have reviewed the history of the Committee, its small budget, its large yield, its integrated set of objectives and, from the viewpoint of science, its highly professional cultivation of scrupulously examined projects and from the viewpoint of the individual researcher, its friendly and socially powerful endorsement by way of the stature and visibility of the Committee as a whole, derived from the integrity of its individual members and the freedom of their chairman, a scientific colleague, to communicate directly with the investigator and all others concerned. We do not have such a group -- anywhere in this domain -- at this time. We have no cynosure in this still overly "sensitive, delicate" and unreasonably provocative area. We need one. Dr. Yerkes, we need you now.

REFERENCES

Aberle, Sophie D. and Corner, George W. Twenty-five years of sex research. Philadelphia: W. B. Saunders Co., 1953.
Beach, Frank (ed.). Sex and Behavior. New York: John Wiley, 1965.
Kinsey, Alfred C., Pomeroy, W. B., Martin, C. E., and Gebhard, P. H. Sexual behavior in the human female. Philadelphia, W. B. Saunders Co., 1953.
Nowlis, V. Critique and discussion. In Money, John (ed). Sex Research: New Developments. New York: Holt, Rinehart and Winston, 1965.
Pomeroy, Wardell B. Dr. Kinsey and the Institute for Sex Research. New York: Harper and Row, 1972.

IN SEARCH OF THE ELUSIVE SEMIOTIC

Karl H. Pribram

Stanford University
Department of Psychology
Stanford, California

INTRODUCTION

The contributions to primatology of the Yerkes Laboratories have been legion. Most of these contributions have
centered on the biological, psychological and social nature of
the chimpanzee - and occasionally the other great apes. There
was a period in the history of the laboratory, however, when
the focus of interest shifted to the primate brain. It was
during this period that I came under the spell of the laboratories. They were then headed by Karl Lashley and staffed by
persons who in due time have become eminent in their own right:
Donald Hebb, Roger Sperry, Josephine Semmes, Edward Evarts,
Kao Liang Chow, Austin Riesen, to name a few. All of us felt
deeply the heritage left to us Robert M. Yerkes and in our research attempted to blend this heritage of behavioral research
with Lashley's genius for asking penetrating questions about
mechanism.

Today I want to address still another focus of interest
that has pervaded the work of the laboratories in the more
recent past. This third focus is man himself. Comparing man
to his nearest relative should provide insights which might
otherwise escape notice. Most ambitious of these comparisons
are those which deal with the chimpanzee's social-cultural
and communicative achievements to which some form of "language"
is central.

The work of the Kellogs, then that of the Hayes with
Vicky (in which I became intimately involved when Lashley retired and I helped guide the laboratories through the next few
difficult years), and now the current studies by Rumbaugh and
his colleagues attests to the continuity of this theme in the
concerns of the Yerkes laboratories. It is this theme which I

* This work was supported by NIMH Grant No. MH12970-08
and NIMH Career Award No. MH15214-13 to the author.

want to elaborate in my presentation – feeling, however, that
I am bringing goals to Newcastle. Project Lana appears to me
to be much more fascinating than what I am about to report.
Nonetheless, since my interests are so parallel to those that
motivate Project Lana, perhaps avenues of productive interchange will be opened by the presentation.

The nineteen-sixties were marked by a great interest in
human language and the structure of its grammar -- an interest
rooted in the conviction that the syntactic organization of
ordinary linguistic communication is similar in its construction to programs used by digital computers (Chomsky, 1957;
Miller, Galanter and Pribram, 1960). More recently, however,
it has become increasingly clear that semiotics -- the study
of the manner in which meanings become linguistically communicated -- must go beyond syntax to semantics if it is to achieve
a coherent view of what occurs when words are used in thinking
and speaking.

I want to take this opportunity to recapitulate and extend an earlier proposal (Pribram, 1973) relating meaning to
grammar in terms of the concepts of information measurement
theory and of mathematical psychology by applying this formulation to data on the organization of brain function. The
basic tenet of the proposal is a simple one: grammar relates
to meaning as partitions relate to the sets which they partition. Its application to human language however is not so
simple. Although considerable progress has been made in our
understanding of the elementary syntactic structures underlying the partitioning process, we are only on the threshold
of comprehending what it is that becomes partitioned. Here we
will attempt to tackle issues such as the distinction between
information, in its strict definitional sense, and meaning;
the nature of distinctive features and poetic connotation;
and the brain organizations which dispose toward context-free,
and those which dispose toward context-sensitive constructions.

VARIETY AND CONSTRAINT IN THE DETERMINATION OF MEANING

Early attempts at applying information theory to human
linguistic communication failed to provide fundamental insights largely due to the fact that the theory itself remained
confused on one basic issue -- the distinction between
uncertainty-reducing and uncertainty-enhancing communications.
Information measurement was based on the ability of a communication to resolve a specified uncertainty established by prior
communications on a set -- a circumscribed domain within which
the communication occurs. The amount of reduction of uncertainty was measured in bits of information, uncertainty describing the complexity of the number of partitions on the set

necessary to specify the organization of that set. A cursory
look at the functions of communication, however, reveals that
all communications do not reduce uncertainty. Some merely repeat prior communications, leaving the uncertainty unaffected,
and some in fact increase it by demonstrating a mismatch, a
prior erroneous partitioning of the communicative domain. The
confusion arose because "errors" were uniformly labelled "information" in the sense defined above as the reduction of
uncertainty.

The original Shannon-Weaver (1949) definition of information proved extremely useful in handling problems of artificial communication systems such as the telephone network.
But Shannon-Weaver "information" has nothing in common with
the demonstration of error or mismatch; the amount of information contained in a message does not depend on the processing
of its errors. A long-distance telephone conversation, for
example, may be interrupted by a periodic beat frequency resulting in errors in interpretation on the part of the receiver -- uncertainty as to intention demanding repetition of
the communication. Such repetitions -- redundant communications -- however contain no additional information. They are
aimed at overcoming errors produced by the form of transmission of the communication. Seven years after the advent of
Shannon-Weaver information theory Ross Ashby (1956) detailed
this distinction in terms of variety and constraint, variety
defined as independence of functioning of parts of the set and
constraints defined as the limits on this independence -- or
dependence amongst parts of the set. Variety is thus measured
as information and constraint as redundancy.

According to the Shannon-Weaver interpretation, variety
and constraint are reciprocals: a bit of information reduces
variety, thus enhancing constraint. The example of the long-
distance telephone conversation, however suggests such a
simple conceptualization of the relationship to be mistaken.
Specification of information and redundancy, to be useful in
human communication, must be sought in other terms. Specifically, in any communication system endowed with memory -- the
ability to compare successive communications -- measures on
variety and constraint refer to wholly different aspects of a
message. In such systems variety entails novelty (see e.g.
Brillouin, 1962), thus coming closer to the ordinary meaning
of information. Further, in such systems with a memory component, constraints operating on a communication deal with
its form -- the structural relationships among communicative
events, among parts of the communicative set. In short, in
such systems variety and constraint operate among relationships between partitions. Only when the partitions produce
complete independence among parts of the set can strict

Shannon-Weaver information theory be applied; when, on the
other hand, the partitions can be shown to be partially dependent, as in the hierarchies or net-like configurations of
computer programs, a theory of _meaning_ -- of relationships --
becomes necessary. The theory of information thus becomes a
special case of the theory of meaning.

The problem can be stated in other terms. A human communication may have a referent. Ordinarily referent communication conveys information. Philosophers have long distinguished, however, between reference and meaning. Meaning goes
beyond reference into use, the use which information conveyed
can have to the sender or receiver. Pierce, for example, has
stated, "We are apt to think that what one means to do and the
meaning of a word are quite unrelated measurings of the word
meaning." (Pierce, 1934; see also Pribram, 1972). He points
out that meaning is _always_ related to doing, the pragmatic, in
some way. In short, meaning relates to the _actions_ of
organisms, actions which have survival value.

The relative dependence among parts of a communication,
and their dependence on use, cannot be ascertained from the
syntactical structure of the communication alone. The relationship among the parts of the set is only trivially given by
the structure of the partioning system. Thus the many popular
examples of meaningless but grammatically correct utterances
("the pillow runs the dog") or sentences with ambiguous
meaning ("they are flying planes"). Meaning is given not by
the fact of partitioning _per se_, but by the dependence among
parts of the set and among successive sets. It must thus derive from some additional property of the parts which defines
their relationship to one another. It is this property which
has proved so elusive.

BRAIN AND SYNTAX

In attempting to come closer to the elusive semiotic
let us first describe, as did linguists of the nineteen-
sixties, some properties of syntax with the hope that in so
doing we will be able to specify more precisely what it is
that is missing. Rather than rely exclusively on an analysis
of human language, however, let us reach into the annals of
comparative behavior and brain function for guidance.

In his address of acceptance of the Chair at Edinburgh,
as well as elsewhere, Vowles (1970a; 1970b) suggested that
phylogeny could perhaps be characterized by the evolution of
a grammar of behavior. The proposal was that whereas inverte-
brates show finite-state Markov-type constructions and verte-
brates have developed phrase-structure type hierarchies, human
behavior, _including_ communicative behavior, can be

distinguished by its transformational components. This sweeping theoretical statement, though perhaps wanting in detail (now gradually being provided by other ethologically oriented scientists such as Boor, 1973) struck a deep responsive chord as, at the time, Peter Reynolds was busy organizing observations on the development of primate play by attempting to specify the rule structure -- the syntax -- characterizing these interactions (Reynolds, in press). The grammar of play, however, is but a specific instance of the plans, programs, or syntax by which behavior in general becomes organized.

The essence of a grammar is its capacity to order sequential dependencies among behaviors, including communicative behaviors. When a behavior reflects only the immediate state of the organism doing the communicating, according to Vowles' proposal that behavior or communication can be thought of as being determined by a Markov type process, very much as a set of dice or the image produced by a kaleidoscope depends only on the configuration of the parts at the moment they cease to be perturbed. When a behavior or communication becomes organized according to some set of rules determining the order in which the parts take place, however, a phrase-structure grammar, to use the terminology of linguistic grammarians, becomes entailed. To the student of comparative animal behavior it is quite obvious that human linguistic communications are not the sole examples of such phrase-structure grammars. A good deal of the concatenation of egg-rolling, mating, or maternal behaviors in birds, for instance, depends on "phrases" of behavior triggering some state in the communicant to whom such behavior is addressed, this state then giving rise to another set of behaviors in turn retriggering a change in state of the original communicant, etc. (see for example Hinde, 1959; 1966). The important point to be made here is that it is never a single communicative act or single behavior that is triggered by such changes of state: an entire sequence is generated. The concept of generative grammar, so popular in current linguistics, is thus applicable to many forms of animal as well as human communication. What is believed to be unique in human communication is both the intentionality of the "triggering" and the communicant's ability to transform the rules which determine sequences. Transformational rules must be imposed upon the more primitive, phrase-structure rules to account for the complexity of human utterances.

Brain research has distinguished two types of rule structures, those which are context-free and those which are context-sensitive. Specific brain mechanisms have been identified for each of these categories. Context-free constructions are ordinarily produced by way of making a sensory discrimination -- visual, auditory, somatosensory or gustatory.

Discriminations allow one to identify objects and events and, ultimately, to name them irrespective of the environmental situation in which they may occur. A rose is a rose is a rose whether it appears one one's lapel or in a garbage pail, in an arrangement or alone in a context-free construction.

About twenty years ago that part of the monkey brain dealing with context-free constructions was discovered (Blum, Chow and Pribram, 1950; Harlow, et al., 1952). For vision this is located in the inferior part of the temporal lobe. For many years lesions of the temporal lobe in man had been known to give rise to visual disturbances; it had been thought, however, that this was due to involvement of Henle's loop, a portion of the optic radiation believed to course round the anterior portion of the temporal horn of the lateral ventricle. When neurosurgeons, then, began performing anterior-temporal lobectomies without producing visual deficits, the existance of Henle's loop was seriously called into question. In fact, visual difficulties in man, especially those resulting from disturbances in the subdominant hemisphere, arise as they do in the monkey from involvement of the temporal cortex itself (Milner, 1971; 1974).

Behavioral analysis of discrimination deficits in monkeys is thus relevant to the present problem. What this analysis reveals is that cortex lying in the posterior portion of the hemisphere, bounded by the projection areas, can be divided into zones, each associated with a primary sensory modality: somesthesis, taste, audition and vision (Pribram, 1969). Disturbances in sensory discrimination are not due to an inability on the part of the animal to distinguish features differentiating two objects. Monkeys who have learned to visually discriminate between an ashtray and a tobacco tin in a simultaneous situation, for instance, though they show a deficit as compared to their normal controls are not able to use this ability in a successive discrimination in which the tin and ashtray are placed in a central location and the animal is required to go right if the ashtray is present and left if the tin is present (Pribram and Mishkin, 1955). Further tests have shown the difficulty to be the monkey's relative inability to utilize distinctive features of the stimulus. The number of such features utilized by operated as compared to control animals has in fact been quantitatively related to the severity of the deficit (Pribram, 1960; Butter, 1968). Neurological analysis of the mechanism involved in this utilization of distinctive features shows that the pathways involved course downward from the temporal cortex into the visual system, as far as the retina intself (Spinelli and Pribram, 1966; 1967).

The temporal cortex may thus be conceived of as a mechanism for generating rules which categorize more primitive

stochastic imaging processes determined by input within the
primary visual system (Pribram, 1974). Such rules would allow
invariances in input to be identified (Pribram, 1960) When
the temporal cortex is removed and environmental context altered, these invariances can no longer be utilized to guide
behavior. Context-free constructions are therefore dependent
upon rules (i.e. phrase-structure rules) generated by the
temporal cortex. The significance of differences in sensory
input is a function of such rules of utilization. Significant
meanings, or signs, are therefore due to context-free, phrase-
structure-type constructions, the mechanism involved being the
generation of such rules by the temporal cortex and their im-
position through efferent control on sensory input.

A great deal of work has also been done on context-
dependent constructions. The basis of poetic connotation,
context-dependent communications in animal behavioral studies
have as their paradigm delayed response or delayed alternation
performance so extensively used in physiological psychology.
The usefulness in these tasks of a particular behavioral act,
or stimulus, depends not on the momentary situation but on
what has gone before: on the context in which performance
occurs. In this instance the context is a temporal one. Here
again the discovery was made some twenty years ago that the
frontal portion of the monkey's brain is involved in the performance of this type of task (Pribram, 1954). Later the
limbic systems and frontal lobe were shown to be anatomically
related (Pribram, 1958) and the limbic portion of the forebrain also implicated in the proper performance of delayed
response or delayed alternation type tasks (Pribram, Wilson
and Connors, 1962; Pribram, Lim, Poppen and Bagshaw, 1966).

Many years ago the delayed response paradigm was modified, thus: instead of showing the animal where a piece of
food might be hidden, interposing a delay, and then asking him
to find it, a token was instead placed in sight of the animal,
removed, and then the animal was asked to locate food where
the token had appeared. This task, the so-called indirect
version of the delayed response problem, was in turn made more
complex until animals were shown to be capable of working for
tokens themselves, useable only at a later occasion for retrieval of food by deposit in a "chimpomat" (Pribram, 1971).
Tokens with use specific to the situation in which they occur
are usually referred to as symbols, symbolic meaning differing
from significant meaning by this very fact of context dependence. The monkeys were thus shown to be capable of a considerable degree of symbolic behavior.

One shortcoming of initial tests used in brain research
to establish significant and symbolic behavioral capabilities
of primates and trace the neural mechanisms involved in such
behavior has been that we have asked animals to communicate

only through some very simple instrumental act. This defi-
ciency was recently overcome in two studies performed with the
chimpanzee. The Gardners (1969), working at the University of
Nevada, taught their chimpanzee Washoe the use of American
Sign Language. They succeeded in constructing a vocabulary of
approximately 150 words by which Washoe could communicate.
Premack, in another experiment at the University of California
at Santa Barbara (1970), developed the token technique with
his chimpanzee, Sarah, until she could eventually communicate
with the trainer by organizing tokens in several orders of
complexity. As might be expected from the context sensitivity
of tokens, Premack found Sarah's behavior to be highly sensitive to changes in training personnel. The meaning of the
tokens seemed to be too dependent on the specifics of the
training situation. Subhuman primates, in short, have been
taught to communicate with both signs and symbols, using both
context-free and context-sensitive constructions.

These studies of brain function have distinguished
several orders of constraint among communicative events, of
"use" of syntax-finite state, significant and symbolic-rele-
vant to understanding the meaning of human communication. It
is customary to think of these orders as being hierarchically
organized. Perhaps this is so. However, the fact that the
limbic and frontal parts of the forebrain are so intimately
related in delayed alternation and delayed response problems,
while posterior cortex seems to deal with discriminations of
every sort, suggests that instead of a trichotomy, as outlined
by Vowles, four fundamental processes may actually be distin-
guishable. When discriminations are involved, the relationship between sign and referent seems to be a straightforward
one: a sign refers to an invariant part of a stochastically
determined kaleidoscopic image. Perhaps there is a similar
relationship between some finite-state-type processes(es) and
the symbolic domain. One of the puzzles plaguing brain research on the frontal lobes and limbic system is that, whereas
delayed alternation is disrupted by lesions of any limbic or
frontal system, delayed response is not (Mishkin and Pribram,
1954). Delayed response behavior thus seems more specific to
the frontal cortex than to the limbic forebrain. Could it be
that the more ubiquitously involved delayed alternation behavior represents a finite-state-type process? If so, what is
the difference between the state determining alternation and
the finite state process involved in discrimination? We have
already noted that referent behavior addresses an invariant
in a communication. It is tempting then to suggest that
alternation addresses some, but not all, variances in a communication: i.e. only those variances which recur with some
discernable regularity. Recurrent regularities are, of

course, ubiquitous in the internal environments of organisms.
They lead to "steady" states characterized by the alternation
of satiety with hunger, thirst, sexual and respiratory need,
and the like. Hence homeostatic, rather than stochastic, properties determine these states.

Basic to homeostatic processes are the spontaneously
recurrent cyclicities of neuronal networks. Circadean rhythms
and other biological clocks derive from such cyclicities and
Pittendrigh (1974) has suggested a description of such rhyth-
micities in terms of systems of mutually coupled oscillators.
In such systems dominant foci of packmakers evolve by virtue
of "entrainment", or capture, of neighboring oscillators into
a single periodicity. Pattee (1971) has constructed a functional model of such a system and developed a set of theoretical views linking this model with linguistic modes of operation. Thus "entrainments" can be viewed as the primitives of
symbolic processes, just as stochastic, Markov mechanisms are
conceived as basic to communication with significant referents.
In such a scheme entrainment deals with recurrent variances,
while stochastic mechanisms serve to process invariances. A
good deal is becoming known about entrainment; the thesis put
forward here should therefore yield readily testable hypothesis. A guide to their formulation can be taken from
Sherrington's classical analysis of spinal cord mechanisms,
in which the difference between the organization of antago-
nistic and allied reflexes can be discerned. Some years ago
an attempt was made to pursue this insight (Pribram, 1960),
and it may be worthwhile to review the distinction in the
light of more recent information.

The difference between stochastic and entrainable systems is that stochastic processes can be mathematically described in finite terms whereas entrainable processes fall into
the domain of infinite algebras. Mathematical learning
theories of the nineteen-fifties and -sixties developed the
potentialities of stochastic processes in great deal (Bush
and Mostellar, 1955; Estes, 1959). More recently G. Spencer
Brown has developed a simple calculus explicating some of the
logical paradoxes occurring in the "infinite", entrainable,
context-dependent domain (1972). The formal properties of
context-dependent processes have also been explored both in
terms of graph structures (Harary, Norman and Cartwright,
1965) and by computer programmers (e.g. Quillian, 1967). But
perhaps the most penetrating insights have come from Shaw's
application of the mathematics of symmetry groups to the problem (Shaw and Wilson, in press). He points out that such
groups are infinite as opposed to finite and thus capable of
handling the persistent puzzle of the generativity of grammars.
Such generativity derives from the fact that an infinite set

"provides a structure for which it is true that a proper sub-
set is equal to the total set". Thus symbols can apply to a
potentially infinite equivalance class of instances.

BRAIN ORGANIZATION AND MEANING

     As we have seen, insights obtained from studies of brain
function address the question of the structure of syntax; they
do not, however, directly address the fundamental issues of
the organization of meaning. In Chomsky's terms, what has so
far been discussed concerns 'surface' rather than 'deep'
structure. In Jakobsen's vocabulary, a fundamental question
still to be explored is the nature of features and what makes
them distinctive.
     Again brain physiology has recently had a good deal to
say about these issues. Units in the nervous system have been
discovered which are sensitive to surprisingly specific features of the envirnoment (e.g., Mountcastle, 1957; Werner,
1973; Evarts, 1967; Hubel and Wiesel, 1962; Barlow and Hill,
1963; Spinelli, Pribram and Bridgeman, 1970). Further, these
features appear to be differentially organized into configurations in each hemisphere of the human brain. Thus after age
seven or thereabouts damage to the right hemisphere in most of
us results primarily in the impairment of spatial relationships while damage to the left hemisphere impairs primarily
temporal relationships, including the linguistic abilities
which are the concern of this paper. In an elegant series of
experiments Sperry (1970) demonstrated the separateness of
these functions in patients whose hemispheres have been seve-
red by sectioning of the corpus callossum -- the major com-
missure which ordinarily connects them.
     These important contributions, however, also pose signi-
ficant problems of interpretation for the neurolinguist. Are
we to search for a unique brain cell for each distinctive
feature of language? If so, do such brain cells respond in-
nately to their respective features, or do they become responsive only through 'experience'? As most feature sensitive
units discovered so far deal with the spatial aspects of
input, how do such feature sensitivities relate to linguistic
structure in a hemisphere supposedly not processing these
features?
     Many of these puzzling questions resolve themselves when
the evidence is looked at from a somewhat different theoretical perspective. The common interpretation that feature sensitive cells serve as "detectors" for their respective features
has been found wanting. Thus Pollen and Taylor (1974), for
instance, have shown that the output of "complex" cells in

the visual cortex (assumed by most to be detectors of lines
of specified length and/or orientation) is not invariant
across all transformations of input other than orientation.
Changes in luminance, line width, number of lines, and their
spacing all influence the final output of such cells. Only a
network of neurons could thus separate their orientation
specificity from that to one of these variables, say to luminance. Several groups of investigators (Pollen and Taylor,
1974; Campbell and Robson, 1968; Glezer, Ivanoff and
Tscherbach, 1973) have shown that a more accurate interpretation is that such cells are sensitive to spatial frequencies,
as opposed to lines of particular length and orientation per
se, and that it is therefore in error to think of them as
simple "line detectors".

The change in interpretation from sensitivity to line
orientation to one of sensitivity for spatial frequency has
major consequences. As discussed elsewhere (Pribram, 1971;
1974; Pribram, Nuwer and Baron, 1974), a spatial frequency
sensitive mechanism allows image reconstruction with a richness and resolution of detail not possible given only outline
feature detection. Perhaps of greater importance, however, is
that fact that spatial frequency analysis of light in the
visual system, just as temporal frequency analysis of sound in
the auditory system, is accomplished wave mechanically and not
in the digital, quantal domain characteristic almost exclusively of the operation of present-day computers. This radi-
cal shift in emphasis allows us to formulate alternative hypotheses as to the organization of deep structure in the brain
and, as a special case, what might be involved in distinguishing features in speech.

Phoneticians have already clarified the fact that distinctive features of spoken language are most readily analyz-
able in terms of wave forms generated by the vocal apparatus -
vocal cords, larynx, oral cavity, tongue and lips. The
Haskins Group for years has been simulating sounds using
spectral techniques (e.g. Liberman, Cooper, Shankweiler and
Studdert-Kennedy, 1967), and one recent study was able to
decompose speech sounds into some six to eight separate components by performing a Fourier analysis taking into account
both spatial and temporal relations (Port, personal communication).

If, in fact, distinctive features of human linguistic
communication may be identified as wave forms, the deep structure of such communications may be found in the wave mechanical
domain. The computer, with its programmable digital information processing capabilities, has been of great service both
in date analysis and as a model of syntactic superficial
structure. Is there not, then, an information processing

system which can serve with equal efficacy as a model (and in due time perhaps also for data analysis) in our search for the elusive semiotic -- the deep, semantic structure of language?

Optical information processing systems are just beginning to be recognized as useful analogues in studies involving the wave mechanical domain. Aside from their image-constructing capabilities they partake in organizations characterized by the distribution of information produced by interference among wave fronts. This distributed aspect of their organization makes them especially attractive to brain scientists who have been puzzling for years over storage via apparent distribution of input over reaches of brain surface resulting in functions strongly resistant to local damage.

Organizations in which information is optically distributed are called holograms (Gabor, 1969; Stroke, 1969). The proposal therefore has been made that spatial and temporal frequency analysis performed by the brain are indicative of holographic-like neural processes (see Pribram, 1966; 1971; in press; Pribram et al, 1974). It must be borne in mind, however, that it is only the organization of the paths taken by light in optically distributed information systems which is intended when we say that holograms serve as a model for neural processing. The energy involved in the latter case is electrical, not photic.

The suggestion to be seriously entertained is that deep structure, in the final analysis, is semantic structure, and that semantic structure derives from a distributed neural organization akin to that found in the holograms of optical information processing systems. Note that deep structure is not synonymous with holographic organization but derived from it. Syntactic structures, as delineated in the earlier parts of this paper, partition -- map -- a holographic, distributed store of information into useful, meaningful organizations for the organism.

Mapping of a distributed, more or less homogeneous matrix into useful organizations is a commonplace of biology. Thus the morphogenetic field becomes organized into useful structures through the action of inducers which derepress the potentialities of the DNA molecules imbedded in those fields. Thom (1972) recently developed a topological mathematics to describe such mappings. The approaches of Pattee (1971) and Shaw (Shaw and Wilson, in press) described in the section on grammar achieve the same end using even more powerful, mathematically distinct techniques. All of these related approaches can be used to define the origin of the distinctive features of language. Each feature would be occasioned by continuous interactions but the ensuing stabilities, the distinctive features per se, would result when interactions gel into non-linearities -- a process Thom terms a "catastrophe".

Perhaps the difference between right and left hemisphere function may best be conceptualized in terms of whether processes leading to image formation or non-linear catastrophic processes come to be emphasized. More likely, however, a simpler distinction based on sensory (e.g. auditory versus visuosomatosensory) and especially motor mode is responsible. Reynolds (in press) has suggested that differential use of the hands by primates necessitated specialization of function of the cerebral hemispheres. Abler (personal communication) has further suggested that when such specialization occurs a unique problem for innervation of midline structures such as the tongue arises. He has experimentally demonstrated that unless one innervation (usually the right in right-handed persons) dominates, conflicting signals from the two hemispheres disrupt function. In short, once hemispheric specialization has occurred, dominance must follow if the midline structures involved in speech are to function harmoniously. And dominance entails some catastrophic-like "decisional" mechanism which more or less stably "takes over" the innervation of the midline.

CONCLUSION

Neurolinguistics in the near future may be able to contribute as richly as has psycholinguistics in the immediate past to classical problems of human language. Definitions of variety and constraint within the framework of information measurement and processing theory can be used to provide the defining properties of grammar and meaning. Comparative behavioral conceptualizations can be invoked to relate many of the details of syntactical structure to brain organization. Three levels and two modes of organization can be identified: significant (context-free) and symbolic (context-dependent) modes each can operate on a transformational, a phrase-structure or a primitive level. The primitive in the significant mode may be stochastic and finite state; in the symbolic mode entrainable, infinitely recurrent regularities are the most likely candidates. Thus infinite as well as finite state mathematics can be fruitfully applied to the fundamental problems of human linguistic communication.

Not only is the concept of distinctive features capable of being analyzed neurologically, but the nature of the elusive deep structure itself related to our current knowledge of characteristics of information storage in the brain. Models of linguistic organization based on both digital computers and optical information processing systems, such as holograms, may be invoked to resolve essential questions. The hope here is that these proposals and outlines will serve to organize the attack against some of the hitherto intransigent problems which continue to plague an otherwise rapidly advancing linguistic enterprise.

REFERENCES

Ashby, W. R. Design for an intelligence-amplifier. In: C. E. Shannon and J. McCarthy, Automata Studies. Princeton, New Jersey: Princeton University Press, 1956, pp. 215-234.

Beer, G. Ethology -- the zoologist's approach to behavior - Parts 1 and 2. Tuatara, Vol. 11: 170-177 (September, 1963) and Vol. 12: 16-39 (March 1964).

Barlow, H. B. and Hill, R. M. Selective sensitivity to direction of movement in ganglion cells of the rabbit retina. Science, 139: 412-414, 1963.

Blum, J. S., Chow, K. L. and Pribram, K. H. A behavioral analysis of the organization of the parieto-temporo-preoccipital cortex. J. Comp. Neuol., 93: 53-100, 1950.

Brillouin, L. Science and Information Theory (second edition). New York: Academic Press, Inc., 1962.

Brown, G. S. Laws of Form. New York: The Julian Press, Inc., 1972.

Bush, R. R. and Mosteller, F. Stochastic Models for Learning. New York: Wiley and Sons, 1955.

Butter, C. M. The effect of discrimination training on pattern equivalence in monkeys with inferotemporal and lateral striate lesions. Neuropsychologia, 6: 27-40, 1968.

Campbell, F. W. and Robson, J. B. Application of Fourier Analysis to the visibility of graftings. J. Physiol., 197: 551-566, 1968.

Chomsky, N. Syntactic Structures. The Hague: Mouton, 1957.

Estes, W. K. The statistical approach to learning theory. In: S. Koch (ed). Psychology: A Study of a Science, Vol. 2. New York: McGraw-Hill, 1959.

Evarts, E. V. Representation of movements and muscles by pyramidal tract neurons of the precentral motor cortex. In: M. D. Yahr and D. P. Purpura (eds.) Neurophysiological Basis of Normal and Abnormal Motor Activities. Hewlett, New York: Raven Press, pp. 215-254, 1967.

Gabor, D. Information processing with coherent light. Optica. Acta, 16: 519-533, 1969.

Gardner, R. A. and Gardner, B. T. Teaching sign language to a chimpanzee. Science, 165: 664-672, 1969.

Glezer, V. D., Ivanoff, V. A. and Tscherbach, T. A. Investi-
    gation of complex and hypercomplex receptive fields of
    visual cortex of the cat as spatial frequency filters.
    Vision Research, 13. 1075-1904, 1973.
Harlow, H. F., Davis, R. T., Settlage, P. H. and Meyer, D. R.
    Analysis of frontal and posterior association syndromes
    in brain-damaged monkeys. J. Comp. Physiol. Psychol.
    45: 419-429, 1952.
Hinde, R. A. Some recent trends in ethology. In: S. Koch
    (ed.) Psychology: A Study of a Science, Vol. 2.
    New York: McGraw-Hill, 1959
Hinde, R. A. Animal Behavior: A Synthesis of Ethology and
    Comparative Psychology. New York: McGraw-Hill, 1966.
Hubel, D. H. and Wiesel, T. N. Receptive fields, binocular
    interaction and functional architecture in the cat's
    visual cortex. J. Physiol., 160: 106-154, 1962.
Liberman, A. M., Cooper, F. S., Shankweiler, D. P. and
    Studdert-Kennedy, M. Perception of the speech code.
    In: K. H. Pribram (ed.). Brain and Behavior, Vol. 4:
    Adaptation. Harmondsworth, Middlesex, England: Penguin
    Books. Ltd., pp. 105-148, 1969
Miller, G. A., Galanter, E. H. and Pribram, K. H. Plans and
    the Structure of Behavior. New York: Henry Holt and
    Co., 1960.
Milner, B. Interhemispheric difference in the localization of
    psychological processes in man. Brit. Med. Bull., 27(3)
    272-277, 1971.
Milner, B. Hemispheric specialization: scope and limits. In:
    F. O. Schmitt and F. G. Worden (eds.). The Neuro-
    sciences Third Study Program. Cambridge, Massachusetts:
    The MIT Press, pp. 75-89, 1974.
Mishkin, M. and Pribram, K. H. Visual discrimination perfor-
    mance following partial ablations of the temporal lobe:
    I. Ventral vs. lateral. J. Comp. Physiol. Psychol.,
    47: 14-20, 1954.
Mountcastle, V. B. Modality and topographic properties of
    single neurons of cat's somatic sensory cortex. J.
    Neurophysiol. 20: 408-434, 1957.
Pattee, H. H. Physical theories of biological coordination.
    Quarterly Reviews of Biophysics, 4, 2 & 3: 255-276,
    1971.
Pierce, C. S. Collected Papers. (Volumes I-IV). Cambridge:
    Harvard University Press, 1934.
Pittendrigh, C. W. Circadian oscillations in cells and the
    circadian organization of multi-cellular systems. In:
    F. O. Schmitt and F. G. Worden (eds.). The Neuro-
    sciences Third Study Program. Cambridge, Mass., The MIT
    Press, 1974, pp. 437-458.

Pollen, D. A. and Taylor, J. H. The striate cortex and the
     spatial analysis of visual space. In: F. O. Schmitt and
     F. G. Worden (eds.). The Neurosciences Third Study Pro-
     gram. Cambridge, Mass., The MIT Press, 1974, pp. 239-
     247.

Port, R. Oral presentation on speech mechanisms and Fourier
     analysis at the University of Minnesota conference on
     Cognition, Perception and Adaptation (August, 1973).

Premack, D. The education of Sarah: A chimp learns the langu-
     age. Psychology Today, 1970(4): 55-58.

Pribram, K. H. Toward a science of neuropsychology (method
     and data). In: R. A. Patton (ed.) Current Trends in
     Psychology and the Behavioral Sciences. Pittsburgh:
     University of Pittsburgh Press, 1954 (Russian transla-
     tion 1964), pp. 115-142.

Pribram, K. H. Comparative neurology and the evolution of
     behavior. In: Roe, Anne and G. G. Simpson (eds.)
     Behavior and Evolution. New Haven: Yale University
     Press, pp. 140-164, 1958.

Pribram, K. H. The intrinsic systems of the forebrain. In:
     Field, J., Magoun, H. W. and Hall, V. E. (eds) Handbook
     of Physiology, Neurophysiology II. Washington: American
     Physiological Society, pp. 1323-1344, 1960.

Pribram, K. H. Some dimensions of remembering: steps toward
     a neuropsychological model of memory. In: J. Gaito (ed)
     Macromolecules and Behavior. New York: Academic Press,
     pp. 165-187, 1966.

Pribram, K. H. The amnestic syndromes: disturbances in
     coding? In: G. A. Talland and M. Waugh (eds.) The
     Psychopathology of Memory. New York: Academic Press,
     pp. 127-157, 1969.

Pribram, K. H. Languages of the Brain: Experimental Paradoxes
     and Principles in Neuropsychology. Englewood Cliffs:
     Prentice-Hall, Inc., 1971.

Pribram, K. H. Neurological notes on knowing. In: J. R. Royce
     and W. W. Rozeboom (eds.) The Second Banff Conference
     on Theoretical Psychology. New York: Gordon and Breach,
     pp. 449-480, 1972.

Pribram, K. H. The comparative psychology of communication:
     the issue of grammar and meaning. In: E. Tobach, H. E.
     Adler and L. L. Adler (eds.) Comparative Psychology at
     Issue. Vol. 223, Annals of the New York Academy of
     Sciences. New York: New York Academy of Sciences, pp.
     135-143, 1973.

Pribram, K. H. How is it that sensing so much we can do so
     little? In: F. O. Schmitt and F. G. Worden (eds.) The
     Neurosciences Third Study Program. Cambridge, Mass.,
     The MIT Press, 1974, pp. 249-261.

Pribram, K. H. Holonomy and structure in the organization of
    perception. In: Proceedings of the Conference on Images,
    Perception and Knowledge. University of Western Ontario,
    May, 1974 in press.
Pribram, K. H., Lim, H., Poppen, R., and Bagshaw, M. H. Limbic
    lesions and the temporal structure of redundancy. J.
    Comp. Physiol. Psychol., 61: 365-373, 1966.
Pribram, K. H. and Mishkin, M. Simultaneous and successive
    visual discrimination by monkeys with inferotemporal
    lesions. J. Comp. Physiol. Psychol., 48: 198-202, 1955.
Pribram, K. H., Nuwer, M. and Baron, R. The holographic hypo-
    thesis of memory structure in brain function and percep-
    tion. In: R. C. Atkinson, D. H. Krantz, R. C. Luce and
    P. Suppes (eds.) Contemporary Developments in Mathema-
    tical Psychology. San Francisco: W. H. Freeman and
    Co., pp. 416-467, 1974.
Pribram, K. H., Wilson, W. A. and Connors, J. The effects of
    lesions of the medial forebrain on alternation behavior
    of rhesus monkeys. Exp. Neurol., 6: 36-47, 1962.
Quillian, M. R. Word concepts: a theory simulation of some
    basic capabilities. Behav. Sci., 12: 410-430, 1967.
Reynolds, P. C. Handedness and the evolution of the primate
    forelimb. Neuropsychologia, in press.
Reynolds, P. C. Play, language and human evolution. In: J.
    Bruner, A. Jolly and K. Sylva (eds.) Play: Its Evolu-
    tion and Development. Baltimore: Penguin Books (in
    press).
Shannon, C. E. and Weaver, W. The Mathematical Theory of
    Communication. Urbana, Illinois: The University of
    Illinois Press, 1949.
Shaw, R. E. and Wilson, B. E. Abstract conceptual knowledge:
    how we know what we know. In: David Klahr (ed.)
    Cognition and Instruction: Tenth Annual Carnegie-
    Mellon Symposium on Information Processing. Hillsdale,
    New Jersey: Lawrence Erlbaum Associates, (in press).
Sperry, R. W. Perception in the absence of the neocortical
    commissures. In: Perception and Its Disorders. Res.
    Publication A.R.N.M.D., Vol. XLVIII, pp. 123-138, 1970.
Spinelli, D. N. and Pribram, K. H. Changes in visual recovery
    functions produced by temporal lobe stimulation in
    monkeys. Electroencept. clin. Neurophysiol., 20: 44-
    49, 1966.
Spinelli, D. N. and Pribram, K. H. Changes in visual recovery
    functions and unit activity produced by frontal and
    temporal cortex stimulation. Electroencept. clin.
    Neurophysiol. 22: 143-149, 1967.
Spinelli, D. N., Pribram, K. H. and Bridgeman, B. Visual re-
    ceptive field organization of single units in the visual
    cortex of monkey. Intern. J. Neurosciences, 1: 57-74, 1970.

Stroke, G. W. *An Introduction to Coherent Optics and Holography*. New York: Academic Press, 1969.

Thom, R. *Stabilite Structurelle et Morphogenese*. Reading, Massachusetts: W. A. Benjamin, Inc., 1972.

Vowles, D. M. *The psychobiology of aggression*. Lecture delivered at the University of Edinburgh. Edinburgh: University Press, 1970a.

Vowles, D. M. Neuroethology, evolution and grammar. In: *Development and Evolution of Behavior*: *Essays in Memory*. T. C. Schneirla, et al., (eds.) San Francisco: W. H. Freeman and Co., 1970b.

Wada, J., Clark, L. and Hamm, I. Asymmetry of temporal and frontal speech zones. (Paper presented to the 10th International Congress of Neurology, Barcelona, Spain, 1973).

Werner, G. Neural information with stimulus feature extractors. In: F. O. Schmitt and F. G. Worden (eds.) *The Neurosciences Third Study Program*. Cambridge, Mass., The MIT Press, 1974, pp. 171-183.

THE EMERGENCE AND STATE OF APE LANGUAGE RESEARCH [1]

Duane M. Rumbaugh

Georgia State University and
Yerkes Regional Primate Research Center
Atlanta, Georgia

In recent years, the premise that man is alone in the animal kingdom in possessing true linguistic skills has been challenged by a number of studies with apes. While it remains true that only man develops a language spontaneously, and that man is the only creature with vocal linguistic skills, the results of ape-language projects over the last decade indicate that the requisites to linguistic competence are not uniquely human (Rumbaugh, 1977).

The success that researchers have had in testing the language acquisition process in apes is a recent breakthrough; however, the theory that man's nearest primate relatives might provide the solution to questions about the nature and evolution of man's linguistic competence is old. In his review of language origin theories, Hewes (1977) brings attention to the conjecture of La Mettrie in 1748 (published in 1912) that it might be possible to teach language to an ape.

La Mettrie indicated that he would choose a young ape, one with the "most intelligent face", and one whose mastery of tasks confirmed that it was indeed intelligent. He would pair the ape with an excellent teacher named Amman, whose ability to teach language to humans born deaf had left a lasting impression. La Mettrie thought that apes might even be taught to speak, and felt that if they could learn to do so, they would then surely know "a language." He viewed language skills as the origin of law, science, and the fine arts, and held that they had served to polish "the rough diamond" of the human mind. Should the ape master language, he, too, then would be gentled; he would no longer be "a wild man, nor a defective man, but he would be a perfect man, a little gentleman..."

Later, at the turn of the current century, experiments by Witmer (1909) and Furness (1916) netted only the slightest

[1] Supported by NIH grants HD-06016 and RR-00165

suggestion of success in teaching apes to speak. Despite arduous effort, their chimpanzee and orangutan subjects mastered
only one or two approximations of words.

In 1925, Robert Yerkes, a pivotal figure in the field of
primate research, also speculated that apes might have capabilities that would allow for the mastery of at least limited
linguistic skills. Though not overly optimistic regarding the
chimpanzee's potential for learning to literally speak, he did
not disallow that possibility. He said, "Although the young
chimpanzees uses significant sounds in considerable number and
variety, it does not, in the ordinary and proper meaning of
the term, speak. Consequently there is no chimpanzee language,
although there certainly is a useful substitute which might
readily be developed or transformed into a true language if
the animals could be induced to imitate sounds persistently."
(Yerkes and Learned, 1925, pg. 60).

In early research conducted in Franklin, New Hampshire,
Yerkes attempted to teach a chimpanzee named Chim to speak. A
small hole in Chim's cage allowed for the delivery of banana
pieces to the chimp as the experimenter made the sounds, "ba,
ba." This was done to attract Chim's attention to the relation between the sound and the preferred fruit, and also to
"stir him to attempt to make the sound on his own account."
After two weeks of twice-daily sessions, both Chim and the
experimenter lost interest in continuing what was deemed an
unsatisfactory method, for Chim had not attempted to reproduce
the sound. Other similar efforts were likewise abandoned.

In 1929, Yerkes concluded, "Evidently, despite possession of a vocal mechanism which closely resembles the human,
and tendency to produce sounds which vary greatly in quality
and intensity, the chimpanzee has surprisingly little tendency
to reproduce other of the sounds which it hears than those
characteristic of the species, and very limited ability to
learn to use new sounds either affectively or ideationally."
(1929, pg. 307). It was surprising to Yerkes that apes did
not put their vocal apparatus to greater use, for he believed
that otherwise their intercommunication is highly complex.
However, Yerkes observed that the apes' vocalizations seemed
to be no more than "innate emotional expressions." He continues, "I am inclined to conclude from various evidences that
the great apes have plenty to talk about, but no gift for the
use of sounds to represent individual, as contrasted with
racial, feelings or ideas. Perhaps they can be taught to use
their fingers, somewhat as does the deaf and dumb person, and
thus helped to acquire a simple, nonvocal 'sign language'"
(1925, pg. 179-180).

Yerkes' observations anticipated later findings
(Lieberman, 1968, 1975) which indicate that it may be physio-

logically and anatomically impossible for apes to produce the
variety of sounds and modulations thereof which constitute the
phones and phonemes of most human languages However, before
this view came to light, several other experiments were undertaken. Kellogg (1968) reports of five extended attempts by
professional psychologists to teach human speech to chimpanzees.
The Kelloggs' own work with a chimp named Gua demonstrated that
the chimpanzee does have a rather well developed ability to
"understand" simple voiced commands. In brief, their chimpanzee appeared to have at least a receptive competence for language if not an expressive one. Gua was not subject to extensive,
systematic training to develop language, but there was nothing
to prevent her from developing an expressive linguistic competence on her own in the enriched home environment in which
she was kept had she been so disposed and able to do so.

The most prolonged and intensive attempt to teach human
speech to a chimpanzee was conducted by Keith and Cathy Hayes
between 1947 and 1954. The Hayes (1971) hypothesized that a
rich vocal environment similar to that experienced by human
infants might be necessary to stimulate language acquisition
by the chimpanzee. Accordingly, they undertook to raise a baby
chimp, Viki, in their own household, giving her all the auditory linguistic benefits of a human child.

Although Viki did not succeed in reproducing more than
four voiceless approximations of "mama," "papa," "cup," and
"up," and the experiment was from that standpoint disappointing
it was fruitful in other respects. It was observed that Viki
linked specific gestures to her utterances as well as using
gestures by themselves to communicate. For example, Viki would
"ask" for a ride by gesturing towards the car outside, by
leading one of the Hayes by the hand to the drawer which contained the key to the front door, by bringing them a purse
always taken along on car rides, and by either showing a picture of a car to the Hayes or by clicking her teeth together
as a "word" for this request.

The observation of this gestural propensity, coupled with
a written suggestion from Hewes (see Hewes, 1977, for an account of their correspondence), led the Hayes to consider the
use of a gestural rather than vocal system of linguistic instruction. Though they did not institute such instruction
themselves, this was one path on which later research found
success. The Hayes' thorough dedication and failure in their
efforts to teach Viki to speak succeeded in dissuading others
to attempt the same. Their experiment serves as support for
the premise of Yerkes' 1925 statement that "an animal which
lacks the tendency to reinstate auditory stimuli--in other
words to imitate sounds--cannot reasonably be expected to
"talk."

The first successful demonstrations of apes' ability to
acquire language skills, making possible two-way communication
between man and ape, were achieved by two different programs
operating simultaneously: one headed by a husband and wife
team, R. Allen and Beatrice Gardner, of the University of
Nevada, the other headed by David Premack, then of the University of California.

In 1966, the Gardners (1971) commenced training a young
chimpanzee, Washoe, in formal language through the use of
Ameslan, the silent sign system widely used among the deaf.
They thereby obviated the limitations of the chimpanzees ability to articulate human speech. The Gardners found quite
early in their work that Washoe was relatively adept in the
learning of new signs and in their appropriate use. They also
observed that Washoe became inclined to chain signs into a
series by the time she had acquired a "vocabulary" of 8-10
signs.

The Gardners report that during their study Washoe's
communications were not restricted to things that she might
want given to her. She also named, for instance, objects or
pictures of things, ostensibly just for the sake of making
the statement. After three years, Washoe had a vocabulary of
85 signs, to which she sometimes spontaneously added signs of
her own, which seemingly were as logical as those coined by
human signers.

Washoe demonstrated a knowledge of classes of words in
relation to other classes, and rarely confused these classes,
such as in her use of pronouns which differentiate persons
from inanimate objects. In one interesting test, Washoe called
various toy animals presented to her for identification "baby,"
thereby indicating her grasp of the concept of size as a distinct category. The Gardners point out that "new usage (of a
sign) showed that the (sign) denoted a concept and was not a
response tied to a particular instance of this concept,"
(1971). The possibility that apes might master syntax was
strongly suggested by their work, and it has been on this
singular point that other ape language projects have emerged,
each with its own methods and programs of training.

Essentially concurrent with the Gardners' program, David
Premack (1971) was attempting a functional analysis of language through studies with his chimpanzee, Sarah. Premack devised a set of words which corresponded to pieces of plastic
of varied form and color. The training procedures employed
were unique in that each plastic "word" was used to inculcate
a specific linguistic function. Sarah mastered the use of 130
terms with a 75-80% reliability in test situations. These
terms included prepositions, the means of indicating negative
and interrogative sentences, of describing dimensional classes

and of expressing the conditional. Premack's strategy has
been based on his belief that every complex linguistic rule can
be broken down into simple units. The definition of these units
and the teaching of them through appropriate methods can produce language competence even in life forms that have no formal,
public language as employed by man.

The LANA Project (Rumbaugh, et al., 1973, 1977) conducted
at the Yerkes Regional Primate Research Center, built upon selected aspects of both the Gardners' and Premack's work. The
first goal of the LANA Project was to develop a technology and
training system that might serve to objectify observations.
Also, by designing a live-in type of language-training system
that could be operational round-the-clock, it was thought that
language training could be implemented in a manner similar to
that in which children learn and refine the various dimensions
of language use.

The application of computer technology made this goal
attainable. A language (von Glasersfeld, 1977), termed "Yerkish"
was devised for work with Lana chimpanzee. Yerkish words were
composed through the combination of nine arbitrary design elements, such as a triangle, a box, and a wavy line. These were
superimposed on each other to create specific words, or "lexigrams." Each lexigram was pictured on its own key on two control panels--one in the experimenter's room and one in Lana's
quarters. Pressing a key causes that key to light up, and
simultaneously causes the lexigram to be displayed on projection
screens in both rooms. Sequences of lexigrams were displayed on
the screens in the order they were pressed, from left to right.
The position of lexigram keys were frequently reassigned to
assure that the association was made with the lexigram rather
than with a key's location. Each lexigram had one and only one
meaning, thereby eliminating the ambiguity of human language,
in which tone and context often serve to make one's meaning
clear.

In this project, it is significant that the order in which
Lana used the lexigrams was important. The appropriate combination of lexigrams was necessary to the fulfillment of any request made by her. For example, Lana learned that pressing the
lexigram keys to spell out "Please machine give ..." could be
completed for a reward, whereas pressing "Please give machine.."
would not be an acceptable request. She had the option of
correcting herself by erasing incorrectly phrased sentences.
Lana also used the erase option in test situations to correct
improper form in communications initiated by the experimenter,
demonstrating her ability to "read" and understand the visual
apparatus (Rumbaugh & Gill, 1977).

To systematize learning and make it possible for Lana to
apply her lessons to new situations, the grammar was designed

so that all lexigrams presented to her fit into specific conceptual categories, such as "physical objects," "human beings," "edible items," "drinkable items," "physical activities," "perceptual activities," etc.

In Yerkish grammatical system, the proper form was always "actor-activity-(object)." Only lexigrams containing certain color-codes and design elements were suitable for use as the subjects or objects of different activities. For example, "Lana drinks juice" was correct, whereas "Lana eats juice" was as unacceptable as "Juice drinks Lana" or "Lana juice drinks." Lana was further able to indicate two types of sentences other than indicative: the imperative (for requests) and the interrogative (for questions). These markers preceded a communication; a period key terminated each communication.

One major benefit of this system over those relying solely on human observation was that all transmissions to and from Lana were automatically recorded by the computer, even at night, so that no portion of Lana's progress was overlooked or undocumented. Furthermore, the computer was an able and objective instructor, free from any bias which could be communicated to the subject. It was programmed to analyze the correctness of all of Lana's lexigram combinations, thus providing her with instantaneous, unequivocal feedback--a vital aid in the learning process--and taking an impossible burden off the experimenter.

It seems fair to conclude certain findings common to all of the major ape-language projects to date, findings which probably will be corroborated by the other projects as they progress: (1) Apes are relatively adept in the learning of words. Their initial, slow progress gives way to ability which frequently allows for word meaning to be learned in a single presentation. Their readiness to add new signals, where meanings bear no necessary relation to the signals' characteristics, cannot be disputed. This is not to conclude, however, that the meanings of words for the apes are exactly the same as those for man; but they are sufficiently similar to extend communication. It is highly improbable that their meanings are identical, for at least to date we cannot discuss with the apes the nuances of word meaning to the end of enhancing accurate communications, as we do among ourselves: Only through formal behavioral tests and fortuitous developments can the apes' word meanings be deduced. (2) Apes readily come to string signs or words together to form sentences. Furthermore, with instruction, they can master the elements of syntax. (3) Apes bring a readiness to extend their language skills, including the use of rules for the structuring of sentences, beyond the specifics of the contexts within which those skills are learned. Were it not for that readiness, the ape-language projects would be nothing but relatively uninteresting demonstrations of rote learning or the mastery of immutable response chains. (4) Apes have an ability to coin labels and to do so in a way that reflects their apparent sensitivity to the salient characteristics of that which they name. Washoe's terming a duck as "water bird" and a brazil nut a "rock berry" are two excellent instances (Fouts, 1974). Lana's terming a Fanta orange soft drink as the "Coke which-is orange (colored)" and an overly ripe banana as the "banana which-is black" and the fruit, orange, as the "apple which-is orange (colored)" are additional relevant examples. Fouts' work in 1974 with another chimpanzee, Lucy, provides still other important examples--citrus fruits were termed "smell fruits," radishes "cry hurt food," and watermelon "drink fruit" and "candy drink." The existence of relatively advanced encoding processes are strongly supported by these ape-produced labels and serve as important evidence that the ape's cognitive processes entail covert functions that are of a linguistic nature.

The central question is, of course, are the ape's productions, taken more comprehensively, linguistic? There is no question whatsoever that they are communications which have an impact upon other animates within the commerce of social contexts. Similarly, there is no reason to question the adaptiveness of their productions in view of their problem-solving effectiveness. But are they tantamount to language use?

Without a good and generally accepted definition of language, the questions cannot be answered definitively. But it is important to note that regardless of the problem in defining language that many are satisfied that the apes' productions are certainly relevant to the language behavior of man. Too many of their productions have been novel and appropriate to the context with its problems to be solved. Too many of them have defined their own context unrelated to the specifics of the events of the moment. But how much is needed to say that the apes is qualitatively the same as the language behavior of man? To ask more and more of the apes and thereby to continue to escalate the criteria for the apes' productions to be called language, as some would do, is without constructive purpose.

The ape-language projects discussed in this paper have grown out of the evolutionary perspective of comparative psychology and the methods of experimental psychology. Were it not for the basic research of these areas, the excitement spawned by Projects Washoe, Sarah, Lana and others might have eluded us These projects have taken place, however, and are continuing. How shall they affect our views?

Clearly, the projects serve to narrow the gap between man and ape. Although the curtain for the opening of communication between man and ape is being drawn still in total silence, the

* PRIMATE VOCALIZATION: AFFECTIVE OR SYMBOLIC?

Peter Marler

Rockefeller University
New York, New York

It is a sobering yet ultimately healthy experience to
re-read the works of great men of science and to discover the
extent to which they anticipated ideas and developments that
we of our generation had fondly thought to be original. Many
biologists have made such discoveries in the writings of
Charles Darwin. As a student of animal behavior I have had
this experience with the works of some of the great German
zoologists such as Heinroth and Lorenz and most recently, in
the writings of Robert Yerkes. A quotation from 1925 bears so
directly on the occasion of the Yerkes Centennial that I would
like to quote it in full. He is speaking about the vocalizations of higher primates with special reference to the chimpanzee.

> "Everything seems to indicate that their vocalizations
> do not constitute true language, in the sense in which
> Boutan uses the term. Apparently the sounds are pri-
> marily innate emotional expressions. This is surpri-
> sing in view of the evidence that they have ideas,
> and may on occasion act with insight. We may not
> safely assume that they have nothing but feelings to
> express, or even that their word-like sounds always
> lack ideational meaning. Perhaps the chief reason
> for the apes' failure to develop speech is the absence
> of a tendency to imitate sounds. Seeing strongly
> stimulates to imitation; but hearing seems to have no
> such effect. I'm inclined to conclude from various

*The author is indebted for thoughtful discussion and
criticism of this paper to Steven Green, Donald Griffin,
William Mason, David Premack and Thomas A. Sebeok. Research
was supported by NSF grant number BNS75-19431.

evidences that the great apes have plenty to talk about, but no gift for the use of sounds to represent individual, as contrasted with racial, feelings or ideas. Perhaps they can be taught to use their fingers, somewhat as does the deaf and dumb person, and thus helped to acquire a simple, nonvocal, 'sign language'." (Yerkes, 1925, pp. 179-180).

The anticipation of the Gardner's work is uncanny and a merger of this train of thought with Yerkes' conviction that chimpanzees do indeed have the capacity for ideation and symbolic thought leads directly to Rumbaugh's work at the Yerkes Center. In this paper I dwell on one particular issue that the Yerkes quotation also introduces, that of the relationship between sound production and emotional expression. He represents a view still wide-spread today, that animal signalling behavior tends to be associated with "affective" processes, in contrast to the "symbolic" processes that are typical of our own species. These two alternatives are usually presented as different in kind. This is the issue I want to re-examine.

Chimpanzees have a vocalization known as food grunting (van Lawick-Goodall, 1968) which I prefer to call "rough grunting" (Marler, 1976, Marler and Tenaza, in press) with which Yerkes was undoubtedly familiar and which many of you will have heard. It is rather specifically associated in the wild with the discovery and eating of a highly preferred food such as a bunch of ripe palm nuts or, if the experimentor intrudes, a bunch of bananas. It is well illustrated in the sound film which Jane Goodall and I prepared on vocalizations of the chimpanzee (Marler and van Lawick-Goodall, 1971).

This call comes closer than any other in the chimpanzee repertoire to qualifying as a name for a limited class of external referents, in this case, preferred foods. Yet it is clear that the vocalizer is aroused, and many would not hesitate to label this act as a manifestation of emotional arousal: not as "symbolic", but "affective".

In making such a judgement, different people use different criteria. One that recurs focusses on the specificity of the relationship between production of the sound and a particular set of referents in this case a large class of objects with the abstract property of being good to eat. The larger the class, the more likely it is to be "affective". I want to re-examine some of the criteria used, beginning with this one.

As a result of the work of Fossey (1972) we have begun to understand the vocal repertoire of the gorilla. I was intrigued to learn that this species has a sound with a very similar acoustical morphology to chimpanzee rough grunting - it is known as "belching." However there is a difference in the circumstances of use. While gorilla "belching" does occur

during eating, it also occurs in a wide variety of other
situations, associated by Fossey with quiet relaxation, either
while resting, or moving slowly through forage and feeding. It
seems to have much more of a contagious quality than "rough
grunting" so that choruses of "belching" may be heard as a
group of gorillas moves through dense undergrowth.

Field studies have revealed that the gorilla has a much
more coherent social group than the chimpanzee (Schaller, 1963,
Fossey, 1974), which is notorious for the fluid fragmentation
and recombination of social units (van Lawick-Goodall, 1968,
Wrangham, 1975). Gorilla "belching" seems to serve an important function not only in communicating the discovery of food
but also in maintaining group coherence. In a sense the referential designata for gorilla "belching" seem to be more
generalized than those for the "rough grunting" of the chimpanzee.

Should the gorilla call be viewed as more "affective"
than that of the chimpanzee? In an evolutionary sense, one
can readily see how a shift of specificity from the relatively
proscribed designata of rough grunting to the larger frame of
reference of the gorilla call may be viewed as adaptive in
light of differences in their social organization. The chimpanzee has no need for a close-range call to mediate coherence
of close-knit groups and does not possess one.

How readily can we apply notions of "arousal" in explai-
ning this difference? In both, vocalizing animals are obviously
aroused, although not highly so. Nevertheless one could probably fit them at different levels of arousal along some single
dimension. The next question is how many dimensions would be
necessary to explain all of the signalling behavior of the
species? Would a single one going from low to high suffice,
or are we compelled to invoke others, perhaps for example one
involving positive affect and the other negative affect?

Consider another case. Struhsaker (1967) discovered a
remarkable array of alarm calls in the vervet monkey (Fig. 1).
The situations in which the four adult male alarm calls are
uttered can be arranged in a series of increasing specificity.
The first two, the "uh" and the "nyow" are evoked by rather
generalized stimuli, and seem to function primarily to alert
others. The third, the "chutter" is associated with a man or
a snake, especially the latter. The fourth, the "threat-alarm-
bark", the only one which is the sole perogative of the adult
male, is typically given on sighting a major predator. It
evokes not approach and mobbing as is the case with a snake,
but rather precipitant flight to the nearest cover.

This series can be interpreted in terms of an increasing
level of arousal, although to provide an effective explanation

we have to allow for rather fine fragmentation of the arousal continuum, with qualitatively different calls being given at different levels. Such an interpretation is less straight forward with the alarm calls of the female vervet moneky, including all but one of the vervet series. The "rraup" and the "chirp", for example, are given in rather specific circumstances and evoke specific and quite contrasting responses, descent from the treetops in one case, and ascent into the treetops in the other. It is not immediately obvious which is associated with a higher level of arousal although careful study might reveal that there are differences. Yet, while continuing to entertain a role for arousal in the programming of these alarm calls, it is here more than anywhere else in vervet vocal behavior, that one is tempted to think of the signals as functioning symbolically, serving a naming function, and designating particular referents.

VERVET MONKEY ALARM CALLS
AFTER STRUHSAKER 1967

	UH!	NYOW!	CHUTTER	RRAUP	THREAT ALARM-BARK	CHIRP
ADULT ♂	yes	yes	yes (rare)	no	yes	no
ADULT AND YOUNG ♀	yes	yes	yes	yes	no	yes
TYPICAL STIMULUS	Minor mammal predator near	Sudden movement of minor predator	Man or venomous snake — but the chutter is structurally different for man and snake	Initial sighting of eagle	Initially and after sighting major predator (leopard, lion, serval, eagle)	After initial sighting of major predator (leopard, eagle)
TYPICAL RESPONSE OF TROOP MEMBERS	Become alert, look to predator	Look to predator, sometimes flee	Approach snake and escort at safe distance	Flee from treetops and open areas into thickets	Attention and then flight to appropriate cover	Flee from thickets and open areas to branches and canopy

Fig. 1. A diagram of alarm calls of the vervet monkey, stimuli evoking them, and responses to others. After Struhsaker 1967 and personal communication.

Even here no alarm call is completely restricted to a single referent. Yet the group of heterogeneous stimulus objects triggering each one also has a set of unifying properties, namely those that constitute a particular class of dangers for which possession of a name, understood by comparisons in such terms, could be of great value to the vervet monkey.

What I am suggesting is that we could think of alarm calls not so much as having large, ill-defined classes of denotations, but rather as representing highly specific classes of dangers, each favoring a particular escape strategy. In this sense they may be viewed as symbolizing such dangers. If they also carry the hallmarks of affective behavior, with simultaneous signs of arousal, perhaps we could think of these as supplementing and enriching the symbolic function rather than excluding it?

There are at least four critical issues of specificity in characterizing symbolic and affective signalling (Fig. 2). We assume symbols to have a certain <u>referential</u> <u>specificity</u>. Thus to the extent that the vervet monkey chutter may be given to a man or to a snake, it fails to satisfy one of the implicit conditions for a symbolic signal. However it might be a mistake to assume that vervet monkeys classify referential events or think about them, in the same terms that we do. Such assumptions may be one of the most serious obstacle to overcome if we are to understand the natural communications of primates.

<u>FEATURES OF "SPECIFICITY"</u>
<u>DIFFERENTIATING</u>
<u>"SYMBOLIC" FROM "AFFECTIVE"</u> SIGNALLING
1. Referential specificity (to the class of objects, events or properties represented)
2. Motor specificity (relation to other responses to the referent)
3. Physiological specificity (association with arousal and emotion)
4. Temporal specificity (coupling to immediate referential stimulation)

Fig. 2. Features of "specificity" differentiating "symbolic" from "affective" signalling.

Another aspect of symbolic signalling is <u>motor</u> specificity. We require at least the potential ability to perform the motor act of signal production separately from the other behavioral processes associated with responding to the particular referent – an alarm call disassociated from acts of fleeing for example, or a food call separated from eating. This in turn relates to the third issue of <u>physiological</u> specificity which is what I think we really have in mind when we consider the relationship between the production of a signal and arousal and the emotions.

For purely symbolic signalling we require disassociation of production from the emotional responses that are normally associated with whatever the animal is signalling about. And this relates in turn to what might be called temporal specificity, the binding in time of perception of a referent and signalling about it. While these events are readily disassociated in our own symbolic behavior they seem to be much more closely tied in animals. Although, as Yerkes reminded us, animals obviously have memories, they do not seem to signal about remembered events, at least not without the provision of some external cue. If rehearsal of some dangerous past experience is in fact impossible for an animal without the physiological correlates and manifestations of terror, one can imagine that out-of-context rehearsal of such events could cause a good deal of confusion. The issue of motor specificity relates to the same fact that the motor acts that reveal to us the animal's emotional state and it's degree of arousal.

What we seem to find in so much of the natural signalling of animals is coherence between the production of signal and the other behavioral and physiological events that normally accompany the perception and response to the set of referential objects, whatever they may be. More than any other feature I believe it is this coherence that leads us to separate affective from purely symbolic signalling, where the degrees of specificity in all three of these domains can be so very much greater.

What is the proper position to take on this coherence? Should it be viewed as a primitive trait, such that imperfections in the physiological design blur the distinctions between responding to an event and signalling about the event? My own view of the matter is different. For one thing, such disassociation can occur in animals, as in play behavior, which typically involves a reorganization of the normal temporal relationships of signals and the other behaviors usually accompanying them, and perception of their usual referents. So it is by no means a physiological impossibility for animals, though it may well be something that in most circumstances animals can well do without, other than in play, where youthful experimentation free of affective accompaniments may

facilitate ontogenetic experimentation. Far from being an
impediment that somehow blunts the effectiveness of animal
signalling behaviors, I would rather view the affective component as a higly sophisticated overlay that supplements the
symbolic function of animal signals. Far from being detri-
mental, it increases the efficiency of rapid unequivocal communication by creating highly redundant signals whose content
is, I suspect, richer than we often suppose. If at times we
ourselves can afford to pare away so much of the affective
accompaniment of our own speech this is to serve very special
communicative requirements that rarely if ever arise in the
natural behavior of animals, even in those as advanced as the
chimpanzee.

It is also important not to underestimate the potential
richness of the information content of affective signalling.
Menzel's recent work clearly points in this direction (Menzel
1974), and one can look further back in time for a similar
conviction, as in this quotation from Norbert Wiener's 1948
book on cybernetics.

"Suppose I find myself in the woods with an intelligent
savage, who cannot speak my language, and whose language
I cannot speak. Even without any code of sign language
common to the two of us, I can learn a great deal from
him. All I need to do is to be alert to those movements
when he is showing the signs of emotion or interest. I
then cast my eyes around, perhaps paying attention to
the direction of his glance, and fix in memory what I
see or hear. It will not be long before I discover the
things which seem important to him, not because he has
communicated them to me by language, but because I myself have observed them."

Thus with no other signalling elements than signs of
arousal and the deictic property of eyes and where they are
looking, such behavior has rich communicative potential. This
theme, that we have underestimated the potential of affective
signalling is echoed by Premack (1975).

"Consider two ways in which you could benefit from
my knowledge of the conditions next door. I could
return and tell you, 'The apples next door are ripe'.
Alternatively, I could come back from next door
chipper and smiling. On still another occasion I
could return and tell you, 'A tiger is next door'.
Alternatively, I could return mute with fright, dis-
closing an ashen face and quaking limbs. The same
dichotomy could be arranged on numerous occasions.
I could say, 'The peaches next door are ripe' or say
nothing and manifest an intermediate amount of positive
affect since I am only moderately fond of peaches.

Likewise, I might report, 'A snake is next door,' also
an intermediate amount of affect since I am less shaken
by snakes than by tigers."

Premack goes on to develop the differences between two
kinds of signalling, referential (= symbolic) and affective
(= excited or aroused), suggesting that information of the
first kind consists of explicit properties of the world next
door while information of the second kind consists of affective
states, that he assumes to be positive or negative and varying
in degree. He gones on,

"Since changes in the affective states are caused by
changes in the conditions next door, the two kinds
of information are obviously related. In the simplest
case we could arrange that exactly the condition referred to in the symbolic communication be the cause
of the affective state."

As Premack indicates, as long as there is some concordance
between the preferences and aversions of communicants then a
remarkable amount of information can be transmitted by an affective system. While he explicitly restricts himself to
"what" rather than "where" one may note, harking back to the
Wiener quotation, that incorporation of a deictic component in
the signal - pointing or looking - not only indicates where,
but also adds a highly specific connotation - one particular
apple tree.

While Weiner reflected on the possibilities of an arousal
system with a single dimension, Premack has implicated at
least two, a dimension of positive affect concerned with attraction, and a dimension of negative affect concerned with
apprehension and avoidance. However even two dimensions seem
inadequate to explain more than the grossest features of the
diversity of signal structure and usage in animals. Again the
issue of specificity intrudes. For example there would be
considerable advantage, teleologically speaking, if a signaller
could give even a general indication of the class of referent
being signalled about by positive affect. Otherwise a hungry
animal, for example, might approach a referent indicated by a
comparison as worthy of positive affect only to discover that
it indicated not food but a social companion or a mate. Much
time and effort would be saved if, at the very least, the
signaller could give some indication as to whether the referent is environmental in nature or social in nature. Environmental designata might include hazards such as predators or
bad weather or something positive such as resources, food and
drink or a safe resting place. Social referents, on the other
hand, again with either positive or negative connotations,
might include a mate or a companion or an infant in need of
care, all worthy of approach, or negative social information

such as an enraged male of the species, to be avoided.

Intriguingly, I find that some analysis of human emotional states use four dimensions that, with some reshuffling, are reminiscent of those that I have mentioned. Figure 3 is modified from the work of Plutchik (1970), a psychiatrist interested in measuring the behavioral biases of emotionally disturbed patients. His major emotional categories, which can also be viewed as communicative systems, are shown in the center of the diagram. In the outer ring I have added some ethological equivalents to those emotional states at the level of ongoing behavior.

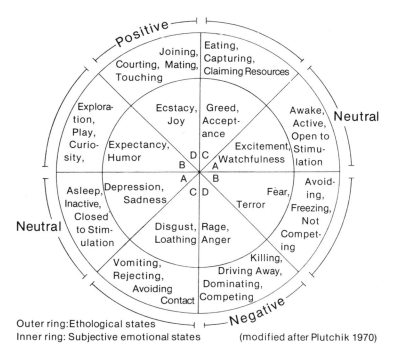

Outer ring: Ethological states
Inner ring: Subjective emotional states   (modified after Plutchik 1970)

Fig. 3. A diagram of human emotional states (inner ring) with some equivalent ethological activities (outer ring) arranged according to whether their connatations are positive, negative or neutral. A-A indicates a possible arousal/depression. B-B a locomotor approach/withdrawal dimension. C-C a social engagement/disengagement dimension. D-D an object acceptance/rejection dimension. Modified after Plutchik 1970.

crude though it is, such a classification encompasses most of
the major sets of activities that comprise the ethogram of an
animal. It is also compatible with the four affective dimensions of arousal/depression (A) approach/withdrawal (B), object
acceptance/rejection (C), and social engagement/disengagement
(D) that I have already mentioned (Fig. 4).

DIMENSION OF AFFECTIVE SIGNALLING

    (a)   Arousal/depression

    (b)   Locomotor approach/withdrawal

    (c)   Object acceptance/rejection

    (d)   Social engagement/disengagement

Fig. 4. Four proposed dimensions of "affective"
signalling.

I am not necessarily wedded to the details of this
particular classification. Rather I am trying in a different
way to make the point that in some communicative circumstances
there are benefits in an increasing degree of specificity in
the multiple relationships between production of a signal, its
underlying physiological basis, and the kind of referential
events that are contingent to its production. I propose that
it may be fruitful to think of symbolic and affective signalling as representing extremes on the same continuum of specificity in these relationships. This would imply that the two
signalling modes are more properly thought of as differing in
degree rather than in kind. The particular constellation of
relationships between referential, motor and physiological
specificity that is manifest in so much animal signalling,
often at the affective extreme, should be taught of not as
something that animals do because they are physiologically incapable of more refined relationships in the direction of
symbolic signalling. Rather it may be viewed as an adaptive
comprise which, even in animals can be seen to move to and
fro, in an evolutionary sense, along a continuum of specificity, as with the gorilla belch and the chimpanzee rough grunt,
depending on what particular solution solves a given behavioral problem most efficiently.

As Yerkes indicated so long ago, and as we now appreciate even better, it is not that animals are by any means incapable of symbolic thought. The point is rather that purely symbolic signalling is a highly specialized mode of operation which, given all the time in the world, has enormous potential, but used in haste in circumstances of dire emergency as is so much more typical of animals, would be hopelessly slow and ambiguous. We see this even in our own behavior, in the intrusion of non-verbal signalling in so many situations that are especially important to us. As everyone ever associated with the name of Yerkes is aware, man has more of the nonhuman primate in him than we often admit.

SUMMARY

At attempt is made to distinguish "affective" from "symbolic" signalling in animals and man. In making such a judgement it is common to appeal to the specificity of the multiple relationships between a signal, its external referents, and associated physiological states and behaviors. "Referential" specificity refers to the class of objects or events represented by the signal. When the class is large we tend to label the signal as affective. Similarly with "motor" specificity, intermingling of signal production with other responses to the referent is viewed as more typical of affective than of symbolic signalling. When "physiological" specificity is broad, so that physiological requirements for signal production are accompanied by signs of arousal and emotion, the signal is usually viewed as affective. With regard to "temporal" specificity, so-called affective signals are usually locked in time to immediate referential stimulation. A review of examples suggests that the degree of specificity exhibited by different animal signals falls on a continuum, so that symbolic and affective signals may be viewed as differing in degree rather than kind. The speculation is offered that in animal signalling, as in human communication, there may be a mingling of symbolic and affective components, the latter having a richer communicative potential than is sometimes supposed.

REFERENCES

Fossey, D. (1972). Vocalizations of the mountain gorilla (Gorilla gorilla beringei). Anim. Behav., 20: 36-53.

Fossey, D. (1974). Observations on the home range of one
    group of mountain gorillas (Gorilla gorilla beringei).
    Anim. Behav., 22: 568-581.
Lawick-Goodall, J. van (1968). The behaviour of free-living
    chimpanzees in the Gombe Stream Reserve. Anim. Behav.
    Monog., 1: 161-311.
Marler, P. (1976). Social organization, communication and
    graded signals: the chimpanzee and the gorilla. In: P.
    Bateson and R. A. Hinde (eds.), Growing Points in Etho-
    logy. Cambridge U. P., England.
Marler, P. and Lawick-Goodall, J. van (1971). Vocalizations
    in Wild Chimpanzees (sound film). Rockefeller University
    Film Service, New York.
Marler, P. and Tenaza, R. (in press). Signalling behavior of
    wild apes with special reference to vocalization. In:
    T. Sebeok (Ed.), How Animals Communicate. Indiana U. P.,
    Bloomington.
Menzel, E. W. (1974). A group of young chimpanzees in a one-
    acre field. In: A. M. Schrier and F. Stollnitz (Eds.),
    Behavior of Nonhuman Primates, Vol. 5. Academic Press,
    New York.
Plutchik, R. (1970). Emotions, evolution, and adaptive
    processes. In: M. B. Arnold (Ed.), Feelings and Emotions.
    Academic Press, New York.
Premack, D. (1975). On the origins of language. In: M. S.
    Gazzaniga and C. B. Blakemore (Eds.), Handbook of Psycho-
    biology. Academic Press, New York.
Schaller, G. B. (1963). The Mountain Gorilla. University of
    Chicago Press, Chicago.
Struhsaker, T. (1967). Auditory communication among vervet
    monkeys, (Cercopithecus aethiops). In: S. A. Altmann (Ed.)
    Social Communication among Primates. University of
    Chicago Press, Chicago.
Wiener, N. (1948). Cybernetics. Wiley and Sons, New York.
Wrangham, R. W. (1975). Behavioural ecology of chimpanzees in
    Gombe National Park, Tanzania. Ph.D. thesis, University
    of Cambridge.
Yerkes, R. M. (1925). Almost Human. The Century Company.

SPONTANEOUS GESTURAL COMMUNICATION AMONG CONSPECIFICS IN THE PYGMY CHIMPANZEE (Pan paniscus)

E. Sue Savage-Rumbaugh; Beverly J. Wilkerson
and Roger Bakeman

Yerkes Regional Primate Research Center,
Emory University and
Georgia State University
Atlanta, Georgia

INTRODUCTION

Recently, a number of studies, such as those in this section, have demonstrated that apes have the ability to acquire skills which are basically linguistic in both form and function. These studies are making a significant contribution to man's understanding of his own language system and the relationship of this system to his cognitive perceptions of the world. They are also adding a new dimension to the question of man's relationship to the ape. Although the biochemical data (Sarich, 1971) had repeatedly indicated that man and ape were far more closely related than were apes and monkeys or even different genera of monkeys, there was an unspoken presumption that these similarities ended with physiology. Even the finding that wild chimpanzees construct and use simple tools and that these skills are transmitted culturally (van Lawick-Goodall, 1968) did not really seem to challenge man's uniqueness. It did serve to remind us that small brained apes can make and use tools, and that a long standing tool use tradition probably preceded any of the cultures represented by the early stone artifacts (Tobias, 1965). The impact of the ape language studies is much broader, however, and is only beginning to be felt.

Menzel (1976) observed that men now view apes with something akin to the wonder and condescending amazement which they formerly accorded primitive peoples. For comparative psychologists, sociobiologists, and ethologists, these studies raise another important evolutionary question: Why is it that the abilities of captive and wild animals to use a complex

linguistic system differ so widely? Could a skill as complex
as that required by even simple linguistic tasks have evolved
far in advance of any semblance of use? Has the chimpanzee
regressed in an evolutionary sense, as Kortlandt and Kooij
(1963) suggest, or are the prerequisites of language to be
found in the natural communicatory system of the chimpanzee,
if we but knew where to look?

Prior to the language studies, it was generally presumed
that apes were not capable of willfully communicating anything
other than basic emotional states (Marler, 1965). Although
there was evidence that apes might be more cognitively oriented
learners than monkeys (Rumbaugh, 1970), the communicative systems of both groups were thought to be similar and basically
primitive if contrasted with man's linguistic system
(Chevalier-Skolnikoff, 1974). The relationship between the
nonverbal forms of communication employed by all other primates and the verbal system of man was, and is, not understood.
Man's facial expressions, his vocal emotive expressions, and
his body postures are usually viewed as having evolved from the
more simplistic expressions and calls of lower primates, and a
distinct line has been drawn between such forms of communication and language because those forms are seen as closed, non-
syntactical, non-symbolic systems (Lancaster, 1975).

Is this distinction a valid one? A brief glance at the
majority of programs used to teach language to severely mentally retarded children indicates that a strong relationship
between verbal and nonverbal communicative skills is believed
to exist by many in this field (Bricker and Bricker, 1974;
Miller and Yoder, 1974). Most programs, in fact, include extensive work in nonverbal communication prior to any form of
verbal instruction; and for the severely retarded, the goal is
merely to induce any sort of two-way communication which will
permit the child to make simple wishes and desires known to
those around him (Graham, 1976).

If two-way nonverbal communicative skills are important
cognitive language prerequisites for retarded human children,
it is reasonable to infer that such skills may also function
as important language prerequisites for the chimpanzees who
are engaged in language learning programs. None of the chimpanzee language programs has made any attempt to teach the animals
two-way nonverbal communicative skills prior to language
training, however, and thus their success would suggest that
the chimpanzee already possesses a high degree of those nonverbal skills which are important to language acquisition.
What are these skills and how are they manifest in spontaneous
inter-animal communication? Can we, through close study of
the chimpanzee's natural communication systems, gain some insight into what these skills might be, how they develop, and

what they imply about the animal's cognitive understanding of the world around it?

The study detailed below is an attempt to investigate one aspect of the spontaneous gestural communication occurring in a group of captive pygmy chimpanzees (Pan paniscus) to the end of determining whether or not they employ, in the absence of training, gestures which are probably learned, nonrandomly displayed, and reliably used to transmit complex types of information.

Although there have been several reports of chimpanzees gesturing in an elaborate manner (Gardner and Gardner, 1971; Hayes and Hayes, 1955; Kellogg, 1968), the reports of inter-animal gesturing indicate that it is an infrequent phenomena and is typically limited to begging, taking, and embracing motions (van Lawick-Goodall, 1968; McGinnis, 1972). Initial observations of a recently captured group of pygmy chimpanzees revealed that these animals regularly gestured to one another, particularly prior to copulation (Savage and Bakeman, 1976). Because such gesturing had (1) not been reported for either wild or captive groups of common chimpanzees (Pan troglodytes), (2) occurred often each day, and (3) appeared to transmit information regarding the ensuing copulatory bout, it was viewed as a potentially fruitful area of study and one which might have implications for the question of language prerequisites.

METHODS

Subjects

Subjects were three wild-caught pygmy chimpanzees (Pan paniscus) housed at the Yerkes Regional Primate Research Center and on lend-lease from the Zairian government. At the initiation of the study, these animals had been in captivity for six months. Sex and approximate ages are given below.

Name	Sex	Age
Lokalema	Female	25 - 35 years
Bosondjo	Male	5 1/2 - 7 1/2 years
Matata	Female	5 1/2 - 7 1/2 years

Procedure

Data were collected during randomly spaced ten-minute sampling periods between 10:00 and 18:00 hours, six days each week from October through April, 1976. The main daily feeding session at 14:00 hours was recorded in entirety. Behaviors were recorded either with a General Electric portable cassette

recorder or a Sony 3640 Videocorder. Video records allowed
for detailed, sequential recording of behaviors and a slow
motion analysis of gestural patterns.

Analyses of gestural data were limited to gestures which
preceded copulatory bouts, so that some nonarbitrary means of
assessing the intent of the gestures could be employed. Nearly
all gestures which preceded copulatory bouts appeared to be
indicators of copulatory position. Pygmy chimpanzees, in contrast to common chimpanzees, employ a wide variety of copulatory positions and, some mutual agreement as to position must be
reached if a completed copulatory bout is to occur (Savage and
Bakeman, 1976). By looking only at gestures which preceded
copulatory bouts and the positional outcomes of each bout, it
was possible to determine whether particular gestures were more
likely to be followed by one type of copulatory position and
other gestures by another. Without some nonarbitrary method
of assessing gestural outcomes, even though the intended
meaning of gestures appears to be clear to the observer, there
is no way to determine whether or not the chimpanzees inter-
pret the gesture in a similar fashion, or even if they place
any significance at all in the gestures.

A total of 21 gestures were specified and described
(see Table I). Analyses are based on 343 observed copulatory
bouts (63 of 67 videotaped and 280 of 672 audio cassette recorded copulatory bouts) that contained at least one gesture,
for a total of 891 gestures. Live observation hours totalled
600.

RESULTS

Types of Gestures

Although the term "gesture" has been used by others to
indicate any specific, stylized body or limb movement (McGinnis,
1972; van Lawick-Goodall, 1968), the use of the term here is
restricted to hand and/or upper forelimb motions. While whole
body postures like "hunched shoulder" are communicative, such
posturings occur throughout the mammalian order, and do not
appear to be prelinguistic forms of communication as do hand
gestures (Hewes, 1973; in press).

Following the identification of 21 individual gestures
and collection of data, gestures were grouped into one of three
broad categories: (1) positioning motions, (2) touch plus iconic
hand motions, and (3) iconic hand motions. These categories
correspond with the presumed evolutionary sequence of the appearance of such gestures.

TABLE I

Gestures observed preceding and during socio-sexual bouts in Pan paniscus grouped into types

1st ORDER GESTURES Positioning Movements	2nd ORDER GESTURES Touch and Iconic Hand and Arm Movements	3rd ORDER GESTURES Iconic Hand and Arm Movements
1. Push limb across body (used to induce partner to turn around)*	1. Touch outside of partner's shoulder, hip or thigh, and motion across body with hand and forearm movement	1. Move hand and forearm across body
2. Push leg or arm out from body; generally performed "en face" (used to move limbs out from partner's venter to facilitate ventro-ventral positioning)*	2. Touch hand or arm and motion outward from partner's body	2. Stand bipedally and wave arms out from body
3. Pull toward self by putting arm around partner's back (used to move partner into proper ventro-ventral position)*		3. Raise arm with palm down
4. Position partner's lower body with both hands (used to induce partner to assume a stance compatible with ventro-dorsal copulation; employed after genitalia of female have already been oriented toward male for ventro-dorsal position)*	4. Rest knuckles on arm or back, and move arm toward self	
5. Pull limb toward self (used to induce partner to move closer to initiator)*	5. Touch shoulder or back and move hand toward self	5. Hold hand toward partner
6. Push under chin (used to induce partner to stand bipedally prior to ventro-ventral standing copulatory bout, or to lie on their backs, prior to a prone ventro-ventral bout)*	6. Touch head, chin or inside of shoulder and lift hand upward	6. Raise arm and flip hand upward at wrist
7. Walk to other end of cage and gaze at partner (used to induce partner to move to another location prior to copulation)*	7. Touch partner and walk to other end of cage	7. Move hand toward another portion of the cage

* Functions, listed in parentheses, are similar across all three gestural types.

101

Positioning movements are the most primitive or basic
type of hand movement and others appear to be derived from
them. They include any use of the hands to move the recipient's
body or limbs in a direction which will promote the assumption
of a particular posture (only positioning movements leading to
copulatory posture are considered here; 1st order gestures in
Table I). During a positioning motion, the initiator usually
held the recipient's limb and gently pushed it in the desired
direction. The actual movement of the limb was done by the
recipient and not the initiator. The initiator merely indicated the desired direction of movement by starting the limb in
that direction.

Touch plus iconic hand motions seems to be a less direct
form of positioning in that the limb or portion of the body to
be moved is lightly touched, then the desired direction of
movement is indicated by the iconic hand motion which depicts
the intended direction. For example, if the initiator wishes
the other animal to turn around, this is so indicated by a
turning motion of the hand at the wrist (2nd order gestures in
Table I).

Completely iconic hand motions form the third category
and, as the heading suggests, the initiator simply indicates,
via an iconic hand movement, what he would like the recipient
to do. These gestures varied from indicating to the recipient
to move his arms out from his body to requesting that he move
his whole body to another location in the cage (3rd order
gestures in Table I).

These basic gestural forms, particular instances of
each, and a typical sequencing of signal exchange prior to
copulation are depicted in Figures 1-10.

Fig. 1.   Bosondjo engages Lokelema in mutual eye contact prior to initiating copulatory signals.

Fig. 2.   Bosondjo touches Lokelema gently on the outside of her shoulder and pushes her upper torso away from himself. This gesture is generally used to induce the recipient to turn around and assume an orientation which will permit dorso-ventral copulation.

Fig. 3.   The female does not turn around, but does display a lips back facial expression, which is often a signal of
copulatory intent.   The male responds by standing bipedally,
touching her lightly on the outside of the other shoulder,
gesturing across his body with his left hand (again, a signal
to move forward past him and assume a dorsal orientation), and
a pursed lips face.

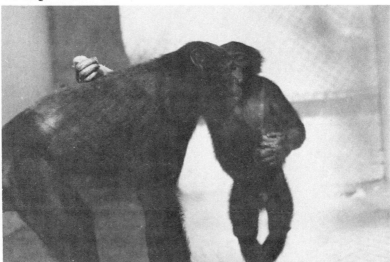

Fig. 4.   The female again ignores his gesture (ignoring
of gestures is common when the recipient prefers to assume a
different posture.   The male moves around to the side and re-
peats his previous gesture.

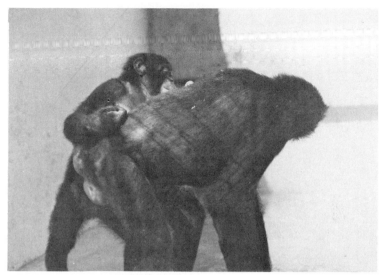

Fig. 5. The female continues to ignore the gestural request of the male, so he moves behind her and attempts to push her into a dorsal orientation. Eye contact is maintained, even though the male is now nearly directly behind the female.

Fig. 6. The female not only ignores the preceding gestures, but turns in the opposite direction so that she is now facing the male.

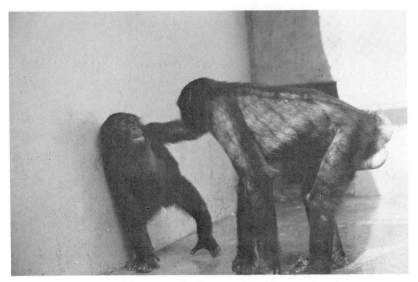

Fig. 7. The male touches her lightly on the inside of the shoulder and lifts his hand upward. This is a gesture which is used to request the recipient to move into a bipedal stance preparatory to ventro-ventral copulation.

Fig. 8. When the female hesitates, the male touches the inside of her arm (a signal to move her arms out from her body) and then rapidly gestures toward his own ventrum.

Fig. 9.   When the female hesitates, the male touches the inside of her arm (a signal to move her arms out from her body) and then rapidly gestures toward his own ventrum.

Fig. 10.   The female approaches and both partners move into a ventro-ventral copulatory bout.

Effects of gesturing

Slow motion analysis of videotaped gestural exchanges strongly suggested that these gestures were not employed randomly. There was often a close correspondence in an iconic sense between gestures and the following body movements of the recipient of the gestures. Coding this type of behavior proved to be extremely difficult, however, as the semantic interpretation of a gesture did not depend on the true form of the gesture per se in a reliable way. Instead, the body orientation of the initiator and recipient determined the exact topography of the gesture in each instance. For example, a hand motion could be interpreted as a signal to approach or to move on past, depending on whether the recipient was already nearby or across the cage. Nevertheless, it was reasoned that the gross body positioning movements preceding ventro-ventral and ventro-dorsal copulatory postures were distinct enough that some gestures could be reliably associated with one posture and others with another. Such a correspondence would probably not be perfect, however, because partners often disagreed on position, and a ventro-ventral signal from one individual would, at times, be followed by a dorso-ventral signal from the partner.

In accordance with the hypothesis that the iconic gestures have grown from or are an evolutionary extension of the touch gestures, which are, themselves, an extension of positioning movements, the frequencies of each such gestural category were summed. Figure 11 displays the relative frequencies of each type of gesture.

Positioning gestures, thought to be most primitive and basic communicators of position, are the most frequent. Completely iconic hand motions, the most highly evolved position indicators, are the least frequent; and the touch gestures fall in-between these two, as would be expected. The preponderance of positioning gestures suggests that this form may still be the most effective method of communication among pygmy chimpanzees.

Analysis of all observed copulatory encounters produced the breakdown shown in Figure 12. Some 40% of all observed copulatory encounters ended in ventro-ventral copulation, 35% ended in ventro-dorsal copulation, and 25% terminated prior to the occurrence of copulation. Gestures were observed in 68% of these bouts; however, this figure underestimates the total number of gestures due to the difficulty of accurately recording the gesture on audio tape. A total of 92% of the bouts recorded on videotape included gestures, since the videotaped gestures could be viewed and reviewed prior to classification.

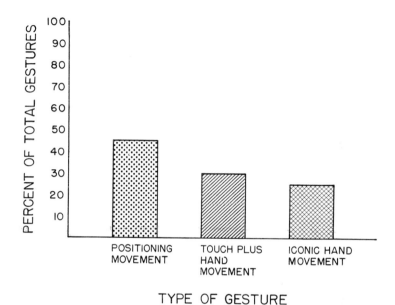

TYPE OF GESTURE

Fig. 11.   Relative frequency of the three forms of gestures displayed preceding and during socio-sexual encounters
in Pan paniscus.

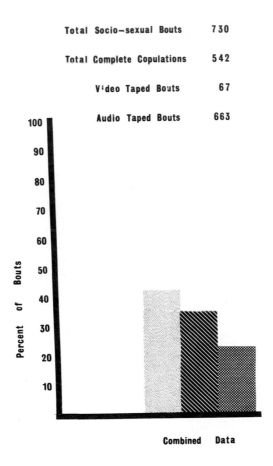

Fig. 12.   Total observed socio-sexual encounters in
Pan paniscus.

The demands of real time audio recording precluded accurate coding of gestures because the majority of gestures occurred so rapidly that reliability was significantly decreased during attempts to record initiator, recipient, gesture, and body orientations of both individuals. The precentage of completed vs. uncompleted ventro-ventral vs. ventro-dorsal bouts, given that one or more gestures were observed prior to copulation, is shown in Figure 13.

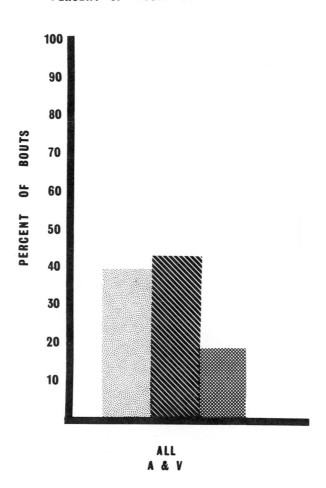

Fig. 13. Socio-sexual bout outcomes in which at least one gesture was observed for both videotaped and audio cassetts recorded data. (Note: see legend in Figure 12).

By comparing Figures 12 and 13, it can be seen that gestures in general increased the overall likelihood that a copulatory bout would be completed (Chi square, significance <.001). Furthermore, while ventro-ventral bouts slightly exceeded ventro-dorsal bouts overall, in those encounters which included one or more gestures, ventro-dorsal bouts exceeded ventro-ventral bouts, suggesting that gesturing may function as an effective means of producing deviations from the normal patterns of behavior.

Breaking down the broad gestural categories into specific types of gestural movements revealed that some gestures were predominantly followed by ventro-ventral copulatory bouts and others ventro-dorsal copulatory bouts (See Figures 14 and 15). Two of the five gestures leading to ventro-dorsal positions

DORSAL   COPULATION

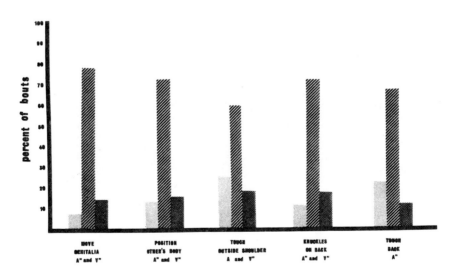

GESTURES

Fig. 14.   Gestures preceding ventro-ventral copulatory bouts. (Note: (1) A indicates that bouts were from videotaped data; (2) ' indicates a Chi square significance <.05, and " indicates a Chi square significance <.01; (3) first bar is ventro-ventral, second is ventro-dorsal, and third bar is no copulation.

were positioning movements.   The others were touch gestures.
One positioning movement, two touch gestures, and two iconic
gestures were significantly associated with ventro-ventral
positions.

VENTRAL   COPULATION

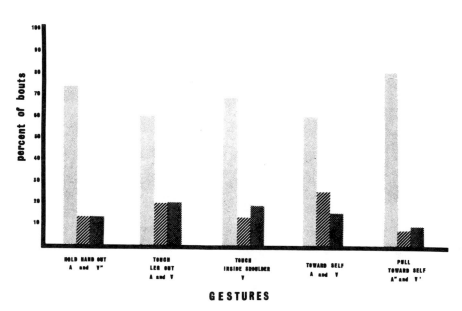

Fig. 15.   Gestures preceding ventro-dorsal copulatory
bouts. (Note: see explanatory notes in Figure 14).

Thus, although gesturing, in general, more frequently
leads to ventro-dorsal positioning, a greater variety of gestural types appears to be associated with ventro-ventral positioning.

DISCUSSION

The gestural propensity and capability of the pygmy
chimpanzee far exceeds that yet reported for any other ape.
This communicatory skill, coupled with the gracile, large
brained skeletal structure reminiscent of early Australopithe-
cines, suggests that Pan paniscus is, perhaps, the best living
animal model to which we may look in the reconstruction of the
history of our own species. The frequent and complex gestural
skills demonstrated by these animals lend strong support to a

gestural origins theory of language evolution (Hewes, 1973; in press). Furthermore, they suggest that inter-animal communication about socio-sexual interactions may have been an important evolutionary pressure in the development of complex cognitively based communicatory skills.

Requesting, via an iconic gesture, that another animal move its body through space to a particular location is not a simple thing. It requires:

    a. a clear concept of self and others,
    b. the realization that personal desires can be communicated to another individual,
    c. that there is a temporary equivalence between the motion of the hand and the movement of the recipient's body,
    d. that the hand is not acting as a hand in the instance of gesturing, but as a symbol for the recipient's body.

These types of symbolic skills are often cited as important language prerequisites (Bricker and Bricker, 1974; Miller and Yoder, 1974), and suggests that pygmy chimpanzees may indeed be important living models for pre-language theory and study.

SUMMARY

(1) Pygmy chimpanzees regularly employ a form of simple gestural communication to reach agreement regarding copulatory position prior to actual copulation.

(2) This gestural system has three basic forms: position movements, touch movements, and iconic hand motions.

(3) This type of regular, complex gestural interchange has not been reported for other apes, and suggests that the pygmy chimpanzee might serve as the best ape model for the study of linguistic precursors and language origins.

REFERENCES

Bricker, W. A. and Bricker, D. D. An early language training strategy. In: R. L. Schiefelbusch and L. L. Lloyd (Eds.), Language Perspectives: Acquisition, Retardation, and Intervention. Baltimore: University Park Press, 1974, pp. 431-468.

Chevalier-Skolnikoff, S. Ontogeny of communication in the stumptail macaque (Macaca arctoides). Contributions to Primatology, Vol. 2. Basel: S. Karger, 1974.

Gardner, B. T. and Gardner, R. A.   Two-way communications with an infant chimpanzee.   In: A. M. Schrier and F. Stollnitz (Eds.), Behavior of Nonhuman Primates, Vol. IV.   New York: Academic Press, 1971, pp. 117-185.

Graham, L. W.   Language programming and intervention.   In: L. L. Lloyd (Ed.), Communication Assessment and Intervention Strategies.   Baltimore: University Park Press, 1976, pp. 371-422.

Hayes, K. J. and Hayes, C.   The cultural capacity of chimpanzee.   In: Nonhuman Primates and Human Evolution.   Wayne State University, 1955.

Hewes, G. W.   Primate communication and the gestural origin of language.   Current Anthropology, 1973, 14, 5-24.

Hewes, G. W.   Current state of the gestural theory of language origins.   Annals of the New York Academy of Science, in press.

Kellogg, W. N.   Communication and language in the home raised chimpanzee.   Science, 1968, 162, 423-427.

Kortlandt, A. and Kooji, M.   Protohominid behavior in primates.   Symposium of the Zoological Society of London, 1963, 10, 61-88.

Lancaster, J. B.   Primate Behavior and the Emergence of Human Culture.   New York: Holt, Rinehart and Winston, 1975.

Marler, P.   Communication in monkeys and apes.   In: I. DeVore (Ed.), Primate Behavior: Field Studies of Monkeys and Apes.   New York: Holt, Rinehart and Winston, 1965, pp. 544-584.

McGinnis, P. K.   Patterns of sexual behavior in a community of free-living chimpanzees.   Unpublished doctoral dissertation, Darwin College, 1972.

Menzel, E.   Implications of chimpanzee language-training experiments for primate field research--and vice versa.   Paper presented at the Sixth International Congress of Primatology, Cambridge, September, 1976.

Miller, J. F. and Yoder, D. E.   An ontogenetic language teaching strategy for retarded children.   In: R. L. Schiefelbusch and L. L. Lloyd (Eds.), Language Perspectives: Acquisitions, Retardation, and Intervention.   Baltimore: University Park Press, 1974, pp. 505-528.

Rumbaugh, D. M.   Learning skills of anthropoids.   In: L. A. Rosenblum (Ed.), Primate Behavior: Development in Field and Laboratory Research, Vol. I.   New York: Academic Press, 1970, pp. 2-70.

Sarich, V. M.   A molecular approach to the question of human origins.   In: P. Dolhinow and V. M. Sarich (Eds.), Background for Man.   Boston: Little, Brown and Company, 1971, pp. 60-81.

Savage, E. S. and Bakeman, R.   Sexual dimorphism and behavior
    in Pan paniscus.   Paper presented at the Sixth Inter-
    national Congress of Primatology, Cambridge, September,
    1976.
Tobias, P. V.   Australopithecus, Homo habilis, tool-using and
    tool making.   South African Archaeological Bulletin, 1965,
    20, 167-192.
van Lawick-Goodall, J.   The behavior of free-living chimpanzees
    in the Gombe Stream Reserve.   Animal Behavior Monograph,
    1968, 1, 161-311.

AMESLAN IN PAN

Roger S. Fouts

University of Oklahoma, Institute of Primate Studies
Department of Psychology
Norman, Oklahoma

Perhaps, to more than anyone scientist, a debt of gratitude is owed to Robert M. Yerkes by those of us who are interested in primate behavior. Robert Yerkes took the label of comparative psychologist seriously, and put into practice the notions that this label implied. He studied not only primates but animals that ranged throughout the phylogenetic scale. It is by comparing the similarities and differences of various animals that we are better able to understand animal behavior in the generic sense. The extreme behavioral and physiological similarities existing between humans and apes have historically aroused the scientific interest of comparative psychologists in studying the behavior of nonhuman primates.

Robert Yerkes was a farsighted individual in terms of noting the cognitive potential of the great apes. In regard to the linguistic abilities of the chimpanzee he stated: "If the imitative tendency of the parrot could be coupled with the quality of intelligence of the chimpanzee, the latter undoubtedly could speak" (Yerkes, 1925, p.53). In 1927 he predicted the future successful scientific attempts to establish two-way communication with a chimpanzee when he stated: "I am inclined to conclude from the various evidences that the great apes have plenty to talk about, but no gift for the use of sounds to represent individual, as contrasted to social, feelings, or ideas. Perhaps they can be taught to use their fingers, somewhat as does the deaf and dumb person, and helped to acquire a simple nonvocal 'sign language'." (Yerkes, 1927, p. 180). In statements such as these and reports, anecdotes and reminiscences of the other speakers at this conference, it is clear that Robert Yerkes demonstrated explicit respect for the great apes as "beings" and for the complexity of their cognitive abilities and potential.

In studying the cognitive behavior of the chimpanzee, Robert Yerkes also demonstrated the importance of establishing a social relationship with the chimpanzee. It is the relationship that is established between two primates - the

117

human and the chimpanzee - that is the key to the success of
the studies examining sign language in the chimpanzee
(The Gardner' (1971) research is the epitome of this). Implicit
in this notion is that the individual establishing the relationship knows the behavioral and biological characteristics
of the species they are working with. Unfortunately, the
naive behaviorism of the 1930's, whose influence is still felt
did not emphasize the fact that animals form relationships
with other animals. Some of this was extreme, for example,
B. F. Skinner initially used to paint his boxes black. He was
apparently more concerned with input-output than he was with
organism inbetween. To ignore the importance of the relationship between animals is to ignore behavior that is most
important to any animal's survival - as an individual or a
species. To put it simply, if an animal cannot establish a
relationship it will not be able to breed and thus procreate
its species. As you examine more complex species the relationship becomes even more important. The long childhood and
adolescence of the chimpanzee is an excellent example of this
importance. It is in the mother-infant relationship that most
of the chimpanzees social learning takes place (van Lawick-
Goodall, 1975). And of course if you are intending to examine
language development the relationship is extremely important
because you need at least two individuals to converse. If
there is no relationship there is no language, since there is
obviously no one to talk with.

     Because of the lack of attention to the relationship in
experimental psychology many experimenters in the past have
used learning paradigms that were developed for rats and
pigeons and then extended this approach beyond the experimental
situation. As a result their experimental results have often
reflected this scientific myopia, rather than truly exploring
the complex mental capacities of chimpanzees.

     As Robert Yerkes (1943) noted one chimpanzee is no
chimpanzee, many experimenters have not taken the gregarious
nature of the chimpanzee into account. The most notable in
the language experiments have been Hayes and Nissen (1971) who
home-reared their chimpanzee Viki for six years and the
Gardners (1971) who home-reared Washoe and are presently
working with three more chimpanzees (Gardner and Gardner, 1975)
These researchers have demonstrated their awareness that
chimpanzees and humans require an adequate environment incor-
porating very close personal relationships with their chimpanzees under conditions that were most favorable to language
acquisition in humans. The Gardners' experiments have taken
into account the chimpanzees' biological and behavioral capacities as evidenced by the fact that they choose a gestural
form of communication which is more suited to the chimpanzee
rather than a vocal form of communication. And all this was

done without comprising scientific procedures and methodology.
The Gardners especially demonstrated this with their use of
the well designed double-blind procedures used in testing
situations.

The naive behaviorism of the 1930's has also affected
the handling of chimpanzees. It has emphasized learning
rather than the biological backgrounds that the subjects bring
to the experimental situation. It also appears to ignore the
fact that organisms develop both physiologically and behaviorally and that the changes associated with these should be
taken into account. As a result some behaviorists have ignored
the needs of the chimpanzee as biological organisms and the
behavioral development of the chimpanzee. This ignorance has
resulted in "lab lore" statements like "you can't work with a
chimpanzee after they reach the age of six." This statement
may be very accurate if you treat a chimpanzee as if it were
a rather large rat or cow. But, if one shows the respect for
the chimpanzee that should be accorded to any higher primate
then it doesn't have to be that way. These "lab lore" statements serve only to threaten a potentially very exciting form
of research - research on language behavior in the mature
chimpanzee. If these types of negative statements are taken
seriously then those scientists who plan to study this behavior throughout the life-span of a chimpanzee may find it
difficult to obtain grants necessary to accomplish this research. Likewise, it makes no sense what so ever to promote
bad public relations for one of the most exciting species a
comparative psychologist has to study. It is my guess that
most probably the rearing conditions, the type of relationship
established and the ignorance on the part of the experimenter
of the biological and behavioral development of the chimpanzee
account for these "lab lore" statements.

For example, when working with children, chimpanzee or
human, it is easy to intimidate them because the experimenter
is bigger than they are. Usually with children their parents
change their dictitorial behavior as the child matures. With
the chimpanzee this may not happen and the surrogate parent's
response to this natural rebellion is permanent caging. As a
result you'll soon find prison behavior appearing in the
chimpanzee, which is not healthy for any primate. Likewise,
if parents continued to treat a 15 year old as if they were
still 5 years old behavior problems would certainly develop
here also.

Of course, the chimpanzee is very strong, but so is the
horse. And like all primates, as the chimpanzee matures he
begins to test the established rules. In humans, if the
parents are understanding, the child forms strong bonds with
his parents and most of the rebellion passes when maturity is

reached, if the child is now respected as an adult. If the parents do not respect these behavioral changes in their child sometimes the child can get locked up too. The same is true for the chimpanzee living in a human society. When the chimpanzee becomes unmanageable, the humans blame it on their chimpanzee rather than accepting the responsibility for their failure as surrogate parents in terms of responding appropriately to the developmental changes in another species.

It has been my experience that the most difficult time in working with chimpanzees is the prepuberty period. It is here that the testing occurs as the chimpanzee attempts to establish his place in the adult hierarchy. This is also a difficult time for humans. Let us look for a moment at what happens to the chimpanzee in the wild. As the young male chimpanzee begins to assert his dominance his mother shows him the proper respect for his new condition. He begins to assert his dominance over another female chimpanzee and begins to leave his mother in favor of male companionship. In the male group he goes through a stressful period of being integrated into the male hierarchy. Nishida (1974), reports that the young male chimpanzee seeks association with the dominant males and as a result he is subjected to frequent attack/submission/reassurance interactions with the males. Nishida states that this ambivalent psychology is probably one of the things that serves to strengthen the male bonds. However, the typical counterpart in captivity often appears to be attack/submission/incarceration. If parents typically locked up their children when they became rebellious then we as a species would not be around very long. As a result reassurance appears to be one of the most important aspects of the human/chimpanzee relationship.

There are people that don't believe the "lab lore" and as a result they are having a great deal of success in home-rearing adult chimpanzees and in maintaining close relationships with adult chimpanzees. Dr. Maurice Temerlin is presently raising a sexually mature chimpanzee in his home. His chimpanzee, Lucy, is now over eleven years old. I am still able to work with Washoe who is now nearly 12 years old and she has been having menstrual cycles for three years. Other "old" chimpanzees are still being handled. Poncho is a 13 to 14 year old male who is still handled on a regular basis. Dr. Sue Savage had much to do with this because of her belief that six years is not the maximum age for working with chimpanzees. It is in this manner that we may be able to study the cognitive and behavioral development throughout the lifespan of our closest living relative. And, this is only possible if we respond appropriately to the biological development of this species.

Now that I've addressed myself to some of the considerations involved in working closely with chimpanzees and the
necessity of establishing proper relationships and the proper
care that is so important to sign language research, I will
now address myself to some of the grammatical aspects of the
use of Ameslan by chimpanzees.

Any research involving a behavior that has been traditionally viewed as being unique to humans seems to cause many
people to look to negative evidence in order to support this
bastion of human uniqueness. One of these is word order or
syntax. It has had a rather contraversial background in regard to Ameslan using chimpanzees. When Washoe learned her
first sign many scientists were found to be incorrect in
assuming that man was the only animal that could label. The
next question raised was in regard to syntax. In an early
diary summary the Gardners reported some combinations that
Washoe had produced. They were not so much concerned with
word order as they were different combinations, otherwise they
would have included the frequency with which the combinations
occurred. As a result the list contains such combinations as
<u>tickle</u> <u>Washoe</u> and <u>Washoe</u> <u>tickle</u>. Some linguists read this
informal report and jumped to the conclusion that the essence
of language was syntax and Washoe didn't have it. This incorrect notion has somehow persisted and it is amusing to say
the least. The argument would be that chimpanzees using sign
language are producing random (non-rule-following) combinations. If this were true it would be a most exciting discovery since as far as I know only computers are able to
approach randomness. To have an organism producing random
(unorganized) behavior is a feat we have yet to find, even in
the remarkable behavior of the chimpanzee. But this is not
an atypical response. I have often felt the same way about
the Spanish language when I have visited Mexico. The Mexicans
certainly weren't using English grammar and in fact to me it
appeared random. But, this was because of <u>my</u> ignorance of
the system rather than the system itself.

Gardner and Gardner (1971) did find evidence for what
they cautiously refer to as "the rudiments of syntax" in
Washoe's signing behavior by noting that her <u>you me action</u>
<u>verb</u> combinations followed a pattern. They collected data
on these combinations for eight weeks. Washoe preferred order
was <u>you me action verb</u>, which was different from the usual way
her human companions combined them: <u>you action verb me</u>. After
six weeks of data collection Washoe changed her preferred
order to match that of her human companions. As a result the
pronoun <u>you</u> preceded the action verb and the pronoun <u>me</u> over
90% of the time. Whereas, because of Washoe's change of
preference, the pronoun <u>me</u> preceded the action verb 60% of the
time and followed it 40% of the time. This was not random

since there was a definite and abrupt change in word order during the sixth week of data collection. Her earlier preferred word order might tell us something about the chimpanzee as a species or at least Washoe as an individual. It is possible that the agents are more important than the action to a chimpanzee.

Fouts, Chown, Kimball, and Couch (1976) have recently completed a study that should put the criticism concerning word order to rest. The study consisted of two experiments: one examining the comprehension of novel phrases and the second examining the active production of novel prepositional phrases. I shall be concerned with only one aspect of the second experiment. But, first I will describe the second experiment. The subject, Ally, was taught to construct sequences of signs describing the relations "on", "in", and "under" between objects arranged before him. He was then tested on the relationship presented to him, using a double-blind procedure, which varied from completely novel (new subject, new locations or a training subject now used as a location with a training location now used as a subject), to a semi-novel (new subject paired with a training location and vice versa) to completely familiar (training relationships).

In training 86 possible arrangements were used, and in testing 240 relationships were used. For the training relationships alone, the prepositions were correct 85% of the time, the locations 78% and the total correct responses 67%. Chance responding is 33% for the prepositions, 16.7% for the locations and 5.6% for the entire relationship. The novel relationships were slightly lower than all of the relationships, but they were still far above chance responding. The prepositions were responded to correctly 77% of the time, location 64% and total 50%. Breaking down the novel relationships into their various categories the results are as follows: for a new location and a new subject the prepositions were 72.2% correct, the locations were 38.9% correct and the total was 33.3% correct; for a new location and a training subject the prepositions were 70% correct, the locations were 40% correct and the total was 23.3% correct; for a new subject and a training location the prepositions were 78.3% correct and total was 62.3%; for training location now used as a subject and vice versa the prepositions were 91.6% correct, the locations were 58.3% correct and the total was 58.3% correct.

Ally was not required to use the subject, however if he did include the subject it was scored against him if he made an error. Out of the 240 relationships Ally included the subject 44 times and was correct on 42 of them.

The point I was addressing earlier that rule-following behavior, in terms of sign order, is important here. Ally made no errors in his sign order. He would mislabel the subject, prepositions or location, but never once did he reverse the sign order.

The results of this experiment indicate that a chimpan-zee can master a simple syntactical system. By correctly describing approximately equal percentages of test and training relationships under test conditions it can be inferred that Ally was in control of a simple and productive grammer. He used this grammar actively and spontaneously in response to objects arranged before him to exemplify relations he had pre-viously learned. Ally was able to describe these relationships by composing sequences of signs of varying degrees of novelty. It is important to note that Ally had to construct each de-scription on his own. There was no restriction on the number of possible answers he could give. As mentioned earlier, all of Ally's errors were errors of naming. The order of signs in his response was always subject, preposition, and location; or no response, preposition and location. It may be that the grammatical relationship was the most salient feature for Ally.

REFERENCES

Fouts, R. S., Chown, Wm., Kimball, G. and Couch, J. 1976, Comprehension and production of American Sign Language by a Chimpanzee (Pan). Paper presented at the XXI International Congress of Psychology in Paris, France, July 18-25, 1976.

Gardner, B. T. and Gardner, R. A. 1971. Two-way communication with an infant chimpanzee. In A. M. Schrier and F. Stollnitz (Eds.), Behavior of Nonhuman Primates. New York: Academic Press, Inc., 4, pg. 117-183.

Gardner, R. A. and Gardner, B.T. 1975, Early Signs of language in child and chimpanzee. Science, 187, 752-753.

van Lawick-Goodall, J. 1975. The Chimpanzee. In V. Goodall (ed.), The Quest of Man, London: Plaidon, pg. 130-169.

Hayes, K. and Nissen, C. 1971. Higher mental functions of a home-raised chimpanzee. In A. M. Schrier and F. Stollnitz, (eds.), Behavior of Nonhuman Primates. New York: Academic Press, 4, pg. 59-115.

Nishida, T. 1974. The Social structure of chimpanzees of the Mahali Mountains. The Behavior of Great Apes. Burg Wartenstein Symposium, No. 62, July 1974.

Yerkes, R. M. 1925. Traits of young chimpanzees. In R. M. Yerkes and B. W. Learned (eds.), Chimpanzee intelligence and its vocal expression. Baltimore: William and Wilkins.

Yerkes, R. M. 1927. Almost Human. New York: The Century Co.

Yerkes, R. M. 1943. Chimpanzees: a laboratory colony. New Haven: Yale University Press.

TALKING TO LANA:
THE QUESTION OF CONVERSATION[1]

Timothy Gill

Yerkes Primate Center and
Georgia State University
Atlanta, Georgia

To what extent can a chimpanzee, with no natural language system of its own, learn an artificial language and use it productively, that is, in conversational problem-solving? This was the question I addressed in recent research with Lana at the Yerkes Center. I designed an experiment in which various misrepresentations of truth and other problem situations would be introduced into Lana's daily feeding routine so as to interfere with her attainment of desired goals. I hoped she would employ her artificial language system in the problem-solving process required of her. The aim of the study was both to test Lana's ability to deal with problem-oriented situations and to discover whether or not Lana actually possessed the ability to converse. To the degree that evidence of these skills could be found in Lana, support would be given for the thesis that humans are not alone in their potential to communicate linguistically.

Lana's success was measured against a working definition of conversation as a linguistic-type of exchange where there is novelty in at least one of the communications transmitted by each of the two beings, and where the topic or subject of the exchange remains relatively organized and constant across time. (Rumbaugh & Gill, 1976).

In order to induce conversation with Lana, an enticement of food or drink was used. Under normal circumstances, Lana's daily feeding consisted of a serving of milk (a preferred liquid) at 9:30 A.M. and a portion of monkey chow (a preferred solid) at 4:30 P.M. Exchanges with Lana were regularly

[1] This research was supported by National Institutes of Health grants NICHD-06016 and RR-00165 to the Yerkes Regional Primate Research Center of Emory University.

initiated by a technician at these times. This routinely
reinforced Lana's language-like behavior, but produced only
brief exchanges which resulted in Lana's machine being stocked
for dispensing the milk or chow. In the morning, the technician would ask Lana by means of his keyboard, ? Lana want what
drink. (Sentence markers precede the sentences in Yerkish,
Lana's language system). Lana would normally answer Lana want
milk. (In the afternoon, Lana was asked ? Lana want what eat,
and chow was requested.)

    To stimulate longer conversations, I devised several
methods of altering Lana's feeding pattern, each causing a delay or diminution of her gratification. As stated above, it
was hoped that the potential frustration generated by these
tactics would spur Lana to employ her language capacity to re-
ctify these situations. For the purpose of this experiment,
only linguistic expressions gained a response; when Lana's
frustration was manifested visually (e.g. whimpering), it had
no effect upon the experimenter's behavior. It was up to Lana
to devise a linguistic solution to fit the problem.

    Lana was given no demonstration nor instruction about
how to make a change in the experimenter's behavior; all linguistic responses reported here were novel productions by her.
Furthermore, in the two years of language training which Lana
had undergone prior to this experiment, she had never encountered similar situations, nor had any experimenter previously
"lied" to her about any conditions, which was one of the
tactics I employed.

    A description of the experiment is presented below. It
took place only at the regular twice-daily feeding times, and
lasted for a period of 16 days.

    The "enticement phase" (Part A) was the first part of
each session, and was initiated by me in one of three ways:
(1) the experimenter (myself) would use his keyboard to ask
? Lana want what drink; or (2) I would ask ? Lana want what
eat; or (3) I would enter the anteroom with either a food or
a drink with which Lana was acquainted and show it to her to
elicit a response (i.e., a request for that item). Only one
of these alternatives was used during a given session. The
order of presenting the alternatives was randomized with the
restriction that each be offered at least 10 times within the
32 sessions.

    These three alternatives were chosen because they were,
at the time of the experiment, the normal means whereby Lana
and I engaged in conversation. Although the third alternative
in particular had already provided the occasion for a number
of exchanges with Lana as reported elsewhere, (Rumbaugh & Gill,
1976), only the first and second alternatives had been utilized extensively with her, and they, in fact, constituted inte-

gral parts of her daily routine. Through the subsequent presentation of each of these three alternatives at random, it
was hoped that the degree to which a normal daily routine
could be utilized as an experimental paradigm would be determined.

Normally, Lana would be asked ? Lana want what drink only
in the morning when she obtained her usual serving of milk,
and likewise would be asked ? Lana want what eat only at other
times during the day, usually at her 4:30 afternoon feeding.
Through the random selection of the three alternatives, however,
it was possible for these questions to be posed at times which
were inappropriate in the context of Lana's established routine. For example, she could be asked ? Lana want what eat in
the morning, when she would normally be asked ? Lana want what
drink as a prelude to receiving her usual milk. It was expected that to the degree that Lana's goal and the question/
situation corresponded to each other, the ensuing conversation
would be relatively brief. To the degree that they differed,
a more prolonged conversation was predicted.

The second phase of the experiment (Part B) encompassed
Lana's initial response, the presence of which was necessary,
by definition, for a conversation to exist. Taken together,
Parts A and B indicate Lana's ability to read the questions
posed to her and to decide upon a linguistic course of action.

The final phase of the experiment (Parts C and D are
omitted, see Gill, 1977, for a complete description of the
experiment) under consideration here (Part E) was that in which
the actual feeding pattern was manipulated by the experimenter.
In Part E, four alternatives were possible. The first (E1) involved stocking Lana's vending device with the item Lana had
requested. The actual food item would be determined by Lana's
response in Part B to Part A, where she had either been asked
what she wanted to eat, what she wanted to drink, or had been
shown an item of food or drink. (If Lana did not respond to
the question asked or the food item shown by requesting that
it be placed in the machine, the food or drink normally associated with that time of day would be given to her.) If Lana
was asked what she wanted to drink in the afternoon and she
responded that she wanted milk, she would be given milk, even
though this was inappropriate to the time of day. In E1 then,
Lana would be given whatever she requested.

Alternatives E2 and E3, unlike E1, were deliberate mis-
representatives on the part of the experimenter. E2 called
for cabbage to be substituted for any solid food Lana might
request, and for water to be the substitute for any liquid requested. Lana would consume both of these substances, and had
names for them, but unlike milk and chow, they were not among
her favorites. The E3 alternative was analogous to E2, except

that the restriction called for water to be substituted for
any solid food, and for cabbage to be substituted for any
liquid.

The fourth alternative (E4) was a modification of the
first alternative in that only a portion of the item requested
or called for in Part A was placed in the dispenser; Lana got
what she wanted, but the whole portion was not made available
for her to comsume. The remainder of the item was positioned
in full view of Lana behind her room in the general area of
the vending devices.

The latter three alternatives, especially E2 and E3,
were expected to prolong the conversation because they would
frustrate Lana, wholly or in part, from obtaining her expressed desire. Conversely, it was predicted that when E1 was the
alternative for a given session and Part A had not called for
any manipulations, Lana would be fully satisfied since her
normal routine would have been maintained; these conversations
were expected to be brief.

Underlying these predictions was the assumption that Lana
would discern the difference between the questions asked of her
and would answer them accordingly. It was not known, however,
what course her responses would take. The results of the experiment were as follows:

If the experiment is divided into two sections, one consisting of those conversations where the questions or situation
was appropriate to the time of day and the other consisting of
the conversations initiated with inappropriate questions, no
significant differences are revealed between the mean number
of exchanges nor the mean duration of the conversations.
Exchanges here refers to a statement/question by Lana or the
experimenter followed by an appropriate response to that ques-
tion/statement by the recipient of it. The duration of the
session refers to the total time remaining in the session in
minutes excluding any time spent by Lana consuming food. In
all instances in which the question or situation was appropriate to the time of day, Lana's initial response was correct,
and in 82% of the sessions starting with inappropriate questions, her initial response was also on target.

It is Lana's responses to the inappropriate questions
which are of greatest interest, because these responses hold
the greatest promise for revealing her linguistic abilities.
Lana was presented with inappropriate questions in Part A on
10 occasions. In three of the first four, Lana answered the
question appropriately with a response which suited it; e.g.
if asked ? Lana want what eat in the morning, she responded
Lana want eat bread. In the seven remaining sessions in which
inappropriate questions were asked, Lana modified the question
each time. In four of these sessions she simply negated the
idea of preforming the inappropriate action suggested by the

experimenter. For example:

Tim: ? Lana want what drink. (This is Part A, in the
          afternoon, her normal time for solid food.)
Lana: No Lana want drink. (Negating the idea of drinking
          in the afternoon.)

In the other three sessions, her responses to the inappropriate questions can be taken as declarations of her real
wish. For example:

Tim: ? Lana want what eat. (This is Part A, in the
              morning, her normal time for milk.
Lana: Lana want drink period., or alternatively,

Tim: ? Lana want what drink. (Part A, in the afternoon,
          her normal time for solid food.)
Lana: Lana want drink (pause) eat chow. (She corrected
          herself in mid-sentence after pausing for seve-
          ral seconds at the point of the incorrect acti-
          vity, drink.)

In Part E, when the food substitution variables came into
play, the results are more interesting. It was possible to
divide the conversations into two groups on the basis of the
Part E reward conditions: (1) Lo probability of being extended
conversations (Conditions E1 and E4); and (2) High probability
of being extended conversations (Conditions E2 and E3). The
statistics bear out this separation: the mean duration of the
High probability sessions was 12.6 minutes, while the Lo probability had a mean duration of only 3.0 minutes ($p < .01$).
        In the sessions in which all of the normal item was
loaded into the vending device the conversation with the experimenter ended there, and Lana shifted to addressing the
computer for that item. When only a part of the item was loaded into the machine (alternative E4), the conversation was
always reestablished for one final exchange: ? You put more
(item) in machine.
        It was expected that when Lana was given a reward diffe-
rent from the normal one, or from what she had requested, the
conversation would be prolonged until the desired item was received. In all but one session, the conversation did in fact
persist until the normal or desired item was obtained. Not
until the last session in which a switch was made (cabbage for
chow) did Lana stop short of obtaining her desired incentive:
She requested Please machine give piece of cabbage, ate a few
bites of cabbage, then stopped working altogether and refused
to answer any questions from the experimenter for over an hour.

It is important to note that in <u>every</u> instance in which
a substitution was made, Lana acknowledged that fact. Lana's
responses to the more disagreeable substitutions took this
acknowledgement a step further: in many cases, Lana demonstrated her ability to use her language system in a productive
manner to deal with novel situations.

In all five instances in which water was substituted for
the normal or requested item, Lana asked that the water be removed. In the first session in which water was substituted
for chow (the E3 alternative), Lana responded as follows:
<u>? You move water out-of machine.</u> I responded <u>Yes,</u> and removed
the water. Lana then asked, <u>? You put chow in machine,</u> to
which I again responded <u>Yes,</u> and the conversation ended with
the chow being placed in the machine. Lana's request that the
water be removed was novel; she had not been trained to ask
that things be removed from various locations. Her only previous use of <u>out-of</u> had been in relation to moving herself
"out-of" her room. On three occasions she also asked that
cabbage be removed from the machine. She chose to eat the
cabbage on the other occasions. (Perhaps eating the cabbage
was preferable to haggling with the experimenter over its
presence.)

A significant grammatical event took place on the sixth
day of the experiment. On this occasion, Lana spontaneously
used a verb tense, the simple past, which she had never been
taught. The setting was as follows: I had placed cabbage in
the machine after Lana had requested juice, though I had responded <u>yes</u> to her request. Lana began by asking for a piece
of cabbage two times. I then asked what was in the machine.
After a couple of false starts, Lana typed out <u>Tim you put
cabbage in machine,</u> (which was indeed time). She followed
this description of an action in the past with a request for
an action in the present tense, <u>? Tim put juice in machine.</u>

Previously Lana had used the verb "put" only as an approximate equivalent for "will you" in requests that items be put
in her vending devices, as in the second request cited above.
The sentence, <u>Tim put cabbage in machine</u> clearly indicates
that she meant to express the past tense, for she had accurately described what I had just done. She later used this
same form again to describe a past event. Lana's spontaneous
use of the past tense is important for it implies that a concept of the past is cognitively present in her. Perhaps in
the future linguistic markers for tense can be employed with
her in order to investigate her perception of time. For example, after I had placed cabbage in the vending device, Lana
asked <u>? You put chow in machine,</u> to which I responded <u>Yes.</u>
This exchange was repeated five times. After the sixth time
Lana asked <u>? Chow in machine,</u> to which I again responded <u>Yes.</u>

Lana was in no way willing to agree with this statement, for
she immediately responded with No chow in machine, which was
indeed true. She had taken the initiative and formed a negation that accurately described the situation. She then responded in a similar fashion five additional times after being
"lied" to.

Another example of novel word useage may be cited. When
asked if she wanted any more cabbage to be placed in the vending device after she had already consumed a considerable
portion, Lana replied: No Lana what more. Normally, "more"
would be used as follows: ? You put more (item) in machine.
These examples of novel sentences are the best evidence we
have of Lana's perception of the world around her.

The results of this experiment disclose an interesting
evolution in the nature of Lana's thinking and the development
of her language skills as a means whereby she has gained control over her environment. As stated above, at first Lana
responded to the questions which were inappropriate to the
time of day by going along with them. After these three
occasions she stopped complying with my apparent intentions.
This change in Lana's strategy occurred after two instances
in which the experimental variables called for water to be
substituted for Lana's original request. She had asked that
the water be removed and her request was granted. From then
on, she resisted going along with the inappropriate questions
in Part A., finally insisting that the established feeding
routine be adhered to.

Lana's own experimentation with her language system
possibly led to her discovery of its potential to control
other's behavior. The concept of disagreeing with the experimenter and having the power to assert her own desire was a
novel one, and took a few trials to discover. As was hoped in
all but a few cases Lana and I remained locked in conversation
until Lana had obtained what she desired, the food or drink
normal to that time of day. The increase in the variety of
Lana's sentences may have been due to her increasing sense of
control, and to her increasing awareness of the possibilities
afforded her through the use of her language system. Her use
of negation tends to support this idea, as do her requests for
the removal of unwanted items from the vending device.

Conversations with Lana, then, is expanding to include
exchanges of greater and greater complexity as she realizes
the potential of her language system. For example, she may
use it to investigate her environment. On one occasion when I
had previously placed water in the milk dispenser, I asked
Lana ? What name this that's in machine. Instead of replying
directly by "guessing" or otherwise responding, Lana proceeded to ask for a sample from the machine by saying, Please

<u>machine give milk</u>, which would give her a small amount of whatever was in the milk dispenser. After tasting it, Lana replied: <u>Water name-of this</u>. This pattern became a regular method of investigation. Examples such as this demonstrate Lana's willingness to utilize her linguistic tools to participate in her environment, and give us valuable insight into her thought processes.

In conclusion, it may be stated that the chimpanzee can develop the capacity to conduct a meaningful exchange of information in a manner which is organized and constant over a period of time. Conversations with Lana in this study were consistent and goal-oriented; she persistently solved problems presented to her through the use of her language system. In her many novel and productive statements, Lana has clearly demonstrated that she is operating in the domain of language, once held exclusive to man.

REFERENCES

Gill, T. V. Conversations with Lana (<u>Pan</u>). In D. M. Rumbaugh (Ed.), <u>Language learning by a chimpanzee</u>: <u>The Lana Project</u>. New York: Academic Press, 1977.
Rumbaugh, D. M. and Gill, T. V. Language and the acquisition of language-type skills by a chimpanzee (<u>Pan</u>). In K. Salzinger (Ed.), <u>Psychology in progress</u>, 270. New York: Annals of the New York Academy of Science, 1976.

MATHEMATICAL CAPABILITIES OF LANA CHIMPANZEE

Gwendolyn B. Dooley
Timothy Gill

Georgia State University and
Yerkes Regional Primate Research Center
Atlanta, Georgia

The "language" of mathematics and ordinary language have several characteristics in common (Lenneberg, 1971; Lenneberg & Long, 1974). Numerals correspond to names in language with both constituting distinctive designations for concepts or entities. Functional or relational words of language (and, same as, or) parallel the symbols for mathematical operations (+ , = , ∩). Sentences are similar to mathematical expressions (7+2-4=5, Set A ∩ Set B) since by combining names and operations, more complex ideas can be expressed or new descriptions made.

One of the most important aspects of both mathematics and language is their relational character. Relating is involved in naming since an association is formed between a person, concept, or thing and a particular designatory name, but often more complex relations are also involved. In language, for example, kinship terms like aunt, brother, and grandmother not only name a certain person, but they also imply both the physical person referrent and the kinship relationship denoted. Adjectives such as tall and dark can only be appreciated once their reference base is known, such as a dark day versus a dark night (Lenneberg & Long, 1974).

Numbers have both names and encompass relations with the relations in mathematics being more exactly defined. Numerals are the names for conceptualized quantities. Each number represents many possible relational compositions, such as 3=1+4-2 or 3=251-248, and every arithmetic phrase can be replaced by a number. The particular number denoted by a mathematical phrase can be precisely determined through knowing the cardinal and ordinal values involved. Mathematical relations are therefore more precisely mapped than linguistic relations.

133

Studies by Hicks (1956), Ferster (1966), Ferster and Hammer (1966), Hayes and Nissen (1971), and Rohles and Devine (1966, 1967) suggested that at least rhesus macaques and chimpanzees in particular possess the ability to learn mathematical-type concepts. In addition to the capacity to form number related concepts, great apes have the capabilities for learning language. Language project experiments by Fouts (1974), Gardner and Gardner (1971), Premack (1971), and Rumbaugh and Gill (1976) have demonstrated that especially chimpanzees among the nonhuman primates possess many language-type abilities including naming, reading and "writing." If the basis for linguistic and mathematical constructs are the same, then it would seem that chimpanzees capable of language skills should also be capable of rudimentary mathematical concepts and operations.

Lana is an adolescent chimpanzee with three years of previous language training. Experiments with Lana suggested that she had the mathematical prerequisites in her past linguistic training. She had learned the following concepts (Rumbaugh & Gill, 1976) which seemed necessary to the training of mathematical skills.

1. Naming of objects and people (Tim, machine, box, shoe);
2. Color-naming of objects (orange, red, purple, etc.);
3. Describing relationships involved in the prepositions in, on, under, and behind in reference to objects or people;
4. Same and different labeling in regard to her various objects being compared to each other.

Lana's knowing that objects and animates had names might be extended to her learning names of relative and later particular numbers. Learning that colors varied across objects was a fairly abstract concept, and one that could possibly be extended to number varying across different sizes of sets. Since Lana knew how to describe prepositional relationships such as in, on, behind, and under (Stahlke, H. F. W., 1976), she might label relative numbers with descriptive terms such as more and less. Lana's training in labeling with same and different was important because she had compared objects and used descriptive labels to designate those relationships. These language-type skills of Lana's and the past number experimentation with primates suggested that Lana could learn the labels and relationships involved in relative number concept formation.

ABILITY TO SELECT THE "MORE" QUANTITY WITH FRUIT LOOP
        RATIOS 1:2 -- 4:5

          When Lana was 5 years old, she was tested to determine
what she knew about number prior to any formal training in
mathematical concepts (Dooley & Gill, 1977). She was presented with two quantities of food bits arranged along lines,
which were eventually equated for length during later stages
of this testing. Lana was presented with 15 ratios or comparisons of two simultaneously presented numbers of food bits,
such that the ratio 3:5 represented 3 food bits versus 5 food
bits. The 15 ratios presented included 1:1, 1:2, 1:3, 1:4,
1:5, 2:2, 2:3, 2:4, 2:5, 3:3, 3:4, 3:5, 4:4, 4:5, and 5:5.
          Lana was able to select (and then eat) the "more" food
bits when compared simultaneously to "less" food bits almost
each time she was presented with ratios of unlike terms (1:4,
3:5, etc.). Table 1 shows that Lana was 94% correct in
selecting the larger set of cereal pieces when unequal quantities were presented, and she only selected the right side 54%
of the time when equal quantities of food bits were presented.
Most of her errors occurred when a difference of one cereal
bit existed between the two quantities offered. Her most
frequent errors occurred with the ratio 4 cereal bits versus
5 cereal bits. That Lana averaged 94% correct in distinguishing the larger of two quantities, even when a difference of
one cereal bit existed between the two sets, showed that she
was extremely accurate at discriminating quantity prior to any
formal training in mathematical concepts. Lana's determination
of larger quantity may have been based on any or all of the
possible cues of number, surface area, mass, etc.

TABLE 1

Proportion of Correct Responses in Selecting the Larger Term in Unequal
    Comparisons or in Selecting the Right Side for Equal Comparisons

1:1	(.68)								
1:2	(.95)	2:2	(.52)						
1:3	(1.00)	2:3	(.82)	3:3	(.45)				
1:4	(.98)	2:4	(1.00)	3:4	(.82)	4:4	(.52)		
1:5	(1.00)	2:5	(1.00)	3:5	(.98)	4:5	(.80)	5:5	(.52)

MORE AND LESS RELATIVE NUMBER LABELING IN REGARD TO RATIOS
        1:2--4:5

This study entailed Lana's ability to master more and
less labeling of two compared sets of variable sized washers
presented in 10 ratios. A ratio was operationally defined as
the set of washers presented as stimuli on two washer display
boards such that the group of 2 washers versus 5 washers was
the ratio 2:5. The 10 ratios presented were 1:2, 1:3, 1:4,
1:5, 2:3, 2:4, 2:5, 3:4, 3:5, and 4:5.

The apparatus used consisted of two washer display boards,
a question and answer board, plastic encased facsimiles of
Lana's lexigrams, and a variety of washers. The washer display
boards measured 30.48 cm on each side, and they had a lightly
marked centerline for placement of washers. A large, vertical
question and answer board presented the pertinent questions
in Lana's lexigrams: ? What more or ? What less. These lexigrams attached to the board as did two possible answer lexigrams
of more and less which Lana could select between and use to
label one display board. Since the washers were in 9 various
diameters of 11 mm, 12.5 mm, 14 mm, 17 mm, 18 mm, 19 mm, 25 mm,
32 mm, and 38 mm and since these were randomly selected from a
bowl full of these washers, relative number labeling instead of
relative area labeling was greatly increased. A board with
lexigrams was used instead of proceeding through Lana's compu-
terized language system since this was a time of transition in
transferring her old system to another facility and in building
her new system.

The gradual process of training in more and less labeling
extended over seven months. Finalization of method required
approximately three months, and Lana needed four additional
months to reach final training criterion. During training and
later testing, Lana sat facing the experimenter with the washer
display boards between them and the question and answer board
to Lana's right side. A trial was initiated with the placement of the washers from Lana's right to left along the centerline across both display boards. The washer groups were presented according to a randomized ratio list with the larger
term in each of the 10 ratios also randomized as to right-left
placement.

The behavioral sequence which Lana followed when making
a response began with her observing which question was posed to
her and with her glancing at the washers on the display boards.
She then touched hands with the experimenter for 5 seconds;
this slowed her response rate and gave her the opportunity to
look between the washer sets several times, if she so chose.
Next, she pointed to the question, selected her answer lexigram, and removed it from the question and answer board. Lana

then labeled one of the two washer amounts and lightly tapped
all the washers in the labeled group, as shown in Figure 1.
When Lana had executed the correct responses in this sequence,
she received a small food reward. She very quickly became
accurate at selecting the appropriate answer lexigram which the
question posed. The correct labeling of the relative numbers
required much more time in training.

Fig. 1. Lana tapping two washers after having correctly
labeled them with <u>more</u>. Lana was requested to compare the
ratio 1:2 in this instance and label the greater number with
<u>more</u>.

        The initial training criterion was met when Lana was
85% correct within a session of 40 trials in answering <u>more</u>
questions. Next, the same criterion was required in <u>less</u>
labeling.   These <u>more</u> and <u>less</u> sessions alternated once
criterion was reached until Lana could meet the 85% criterion
within one session for <u>more</u> and one session for <u>less</u>. A final
within-session criterion of 70% correct responses was required
in a session of 40 trials in which blocks of 5 <u>more</u> and 5 <u>less</u>
questions alternated. When Lana reached this last criterion,

testing began with the following session. Once the method was
finalized, Lana reached criterion after 2230 trials with more
labeling and 2150 trials of less labeling. Essentially no
difference existed in labeling ability contrary to some child
development literature (Donaldson & Balfour, 1968; Palermo,
1973, 1974).

Lana's comprehension of relative number labeling with
more and less was tested over 100 trials following the same
basic training procedures except more and less were randomly
presented with the restriction that not more than 3 trials of
either question could be presented. Testing was conducted in
two consecutive sessions of 50 trials each. Lana averaged 89%
correct on more and less labeling. As Table 2 shows, Lana
made the most errors on the ratios with a difference of one
(2:3, 3:4, 4:5) and on the ratios with larger groups of
washers (3:5, 4:5). Lana was 90% correct on more labeling and
88% correct on labeling with less. Thus performance was
essentially the same for both more and less labeling, and she
appeared to be using number as the relevent cue since the
washers always varied in size.

TABLE 2

Proportion of Correct Responses in More and Less
Labeling of Ratios 1:2 - 4:5

A. Composite of Proportion of Correct Responses for both
    More and Less Labeling

1:2 (1.00)			
1:3 (1.00)	2:3 (.90)		
1:4 (1.00)	2:4 (.90)	3:4 (.90)	
1:5 (1.00)	2:5 (1.00)	3:5 (.60)	4:5 (.60)

B. Proportion of Correct Responses for More

1:2 (1.00)			
1:3 (1.00)	2:3 (1.00)		
1:4 (1.00)	2:4 (.80)	3:4 (1.00)	
1:5 (1.00)	2:5 (1.00)	3:5 (.40)	4:5 (.80)

C. Proportion of Correct Responses for Less

1:2 (1.00)			
1:3 (1.00)	2:3 (.80)		
1:4 (1.00)	2:4 (1.00)	3:4 (.80)	
1:5 (1.00)	2:5 (1.00)	3:5 (.80)	4:5 (.40)

To confirm that number was indeed the cue which Lana was using in her _more_ and _less_ labeling, a battery of control tests were presented. The first control test involved equating placement timing of the washers on the two different boards. "More" versus "less" placement time of the numbers of washers to their respective boards might have been a cue for Lana. A large box screen was placed over the washer display boards prior to washer placement to shield Lana's observation of the washers being positioned. The experimenter kept her hand under each side of the box for the same length of time. As a second precaution the question was not completed with _more_ or _less_ until after washer placement and screen removal. On this test Lana was 80% correct. Time taken in washer placement was not cuing Lana in relative number labeling.

A second test controlled for surface area covered by the washers being the relevant cue that Lana might have used in her discriminations. The surface area of each of the nine diameters of washers was calculated eliminating the area of the washer's central "hole". The 10 number ratios were then grouped so that equal surface area ratios ($1 \pm .02$) existed between the two groups of washers in each ratio set. For example, one 32 mm diameter washer of 785 mm^2 was essentially equal in surface area to five washers, one 18 mm, one 19 mm, two 12.5 mm, and one 14 mm, which covered 776 mm^2 ; this resulted in a surface area ratio of 1.01 for this particular number ratio of 1:5. This test was presented to eliminate the possible cue of surface area which had only been randomly varied in previous tests. Testing incorporated two sessions of 50 trials each for a total of 100 trials.

On this crucial test, Lana averaged 76% correct in labeling with _more_ and _less_. She was 86% correct when labeling with _more_ and 67% correct in labeling with _less_. When surface area was held constant and only number varied, Lana again performed well above chance level for both _more_ and _less_ labeling. Lana had learned to use number as the only consistent relevant cue in making her discriminations.

A third control test battery involved an extension of the number of washers used to 10 washers. This test was designed to ascertain Lana's ability to transfer her trained relative number labeling skills to larger numbers of washers and to new configurations of washers. The 45 ratios used in this test were composed of the numbers between 1 and 10 paired only once and eliminating the ratios of 1:1, 2:2, . . . 10:10. This experiment consisted of three parts. Lana was required to label larger numbers of washers arranged in lines as before, she was tested on labeling the same number ratios grouped in cluster versus cluster, and lastly she was required to label the 45 ratios arranged as cluster versus line. A cluster

was placed in either an oval or circle configuration with not
more than 15-20 mm ever separating any washer. All other procedures were the same as in previous training and testing.

On this third control test sequence, Lana's high performance continued in distinguishing and correctly labeling relative numbers of washers with either more or less. When the 45
ratios were compared line to line, Lana averaged 81% correct
with a score of 80% correct on more and 82% accurate on less
labeling. When presented with the larger numbers in cluster
versus cluster, Lana was 87% correct on more labeling and 87%
correct on labeling with less. When cluster was compared to
line with the 45 ratios, Lana averaged 84% correct with a score
of 100% correct with more and 69% correct with less. Larger
numbers and different configurations did not disrupt Lana's
ability to accurately discriminate between two sets of numbers
and then to correctly label either of those sets with either
more or with less. She definitely had the skill of relative
number labeling.

CONCLUSIONS

Prior to formal training in number, Lana had developed
on her own the basis of the relative number concept--she was
94% correct in selecting the "more" cereal bits when compared
to "less" bits when given 400 trials of a pretraining test.
Lana might have developed this number related concept by comparing a large, whole sweet potato to a small cube of sweet
potato or by contrasting a handful of M&Ms to a single M&M,
for example. She could have acquired this continous and/or
discrete quantity concept due to her special language training
emphasing particular qualities of various objects or perhaps
this was due to her "enriched" living conditions; but this is
not necessarily so. Given an enriched environment full of
possible comparisons, maybe many animals can distinguish "more"
food over "less" food.

Following relative number labeling training, Lana's
acquired ability to label varying numbers in 10 different
ratios with either more or less was quite remarkable. She could
label number when surface area was equated between the two sets
in any ratio, when placement timing of the different numbers of
washers was held constant, and when larger numbers of washers
and different configurations of washer numbers were introduced.
That Lana was able to perform at very high levels of accuracy
in relative number labeling in this task demonstrated one of
the clearest examples among nonhuman primates of the ability to
make relational responses based on abstract, contingent cues.
To emphasize how variable the labeling of a particular number

could be, Lana could see 3 washers in either the ratio 1:3 or 3:5. Sometimes 3 would be correct or sometimes incorrect, depending on whether she was asked to label the ratio presented with more or less respectively.

This more and less labeling task probably involved the most abstract concept formation yet required of Lana. Since Lana was an average, laboratory born chimpanzee prior to her linguistic and mathematical training, only her training has set her apart from most other great apes. Probably most other hominoids could learn the same task given Lana's prior training or an enriched environment. What is important is that the ability of a chimpanzee to learn the numerical/mathematical concepts of more and less could be taken as further support for the thesis that the differences between human and chimpanzee is one of degree and not of kind.

REFERENCES

Dooley, G. B., & Gill, T. V. Acquisition and use of mathematical skills by a linguistic chimpanzee. In D. M. Rumbaugh (ed.), Language learning by a chimpanzee: The Lana Project. New York: Academic Press, 1977.

Donaldson, M. & Balfour, G. Less is more: A study of language comprehension in children. British Journal of Psychology, 1968, 59, 461-471.

Ferster, C. B. Arithmetic behavior in chimpanzees. Scientific American, 210, May 1966, 98-106.

Ferster, C. B. & Hammer, C. E. Synthesizing the components of arithmetic behavior. In W. K. Honig (ed.), Operant Behavior. New York: Meredith, 1966.

Fouts, R. S. Language: Origins, definitions, and chimpanzees. Journal of Human Evolution, 3(6), 1974, 475-482.

Gardner, B. T. & Gardner, R. A. Two-way communication with an infant chimpanzee. In A. M. Schrier & F. Stollnitz (eds.) Behavior of nonhuman primates (Vol. 4). New York: Academic Press, 1971.

Hayes, K. J. & Nissen, C. H. Higher mental functions of a home-raised chimpanzee. In A. M. Schrier & F. Stollnitz (eds.), Behavior of nonhuman primates, (Vol. 4). New York: Academic Press, 1971.

Hicks, L. H. An analysis of number-concept formation in the rhesus monkey. J. of Comparative Physiological Psychology, 49, 1956, 212-218.

Lenneberg, E. H. Of language knowledge, apes, and brains. J. of Psycholinguistic Research, 1(1), 1971, 1-29.

Lenneberg, E. H. & Long, B. S. Language development. In J. A. Swets & L. L. Elliot (eds.), Psychology and the handicapped child, Washington, D. C. Govt. Pri. Ofc. 1974.

Palermo, D. S.   More about less: A study of language comprehension.   J. of Verbal Learning and Verbal Behavior, 12, 1973, 211-221.

Palermo, D. S.   Still more about the comprehension of "less". Developmental Psychology, 10, 1974, 827-829.

Premack, D.   On the assessment of language competence in the chimpanzee.   In A. M. Schrier & F. Stollnitz (eds.), Behavior of nonhuman primates, (Vol. 4).   New York: Academic Press, 1971.

Rohles, F. H. & Devine, J. V.   Chimpanzee performance on a problem involving the concept of middleness.   Animal Behavior, 14(1), 1966, 159-162.

Rohles, F. H. & Devine, J. V.   Further studies of the middleness concept with the chimpanzee.   Animal Behavior, 15(1) 1967, 107-112.

Rumbaugh, D. M. & Gill, T. V.   Language and the acquisition of language-type skills by a chimpanzee (Pan).   In K. Salzinger (ed). Psychology in progress: An interim report (Vol. 270).   New York: Annals of the New York Academy of Science, 1976.

Stahlke, H. F. W.   The requisition and use of relational terms in a primate language analogue.   Unpublished manuscript, Georgia State University, 1976.

OBJECT- AND COLOR-NAMING SKILLS OF LANA CHIMPANZEE[1,2]

Susan M. Essock[3], Timothy V. Gill,
and Duane M. Rumbaugh

Yerkes Regional Primate Research Center
Georgia State University
Atlanta, Georgia

When two people communicate, the fact that the speaker can direct the listener's attention to a specific object without having to point to the referenced object is of enormous value. The foundation for such an ability is the existence of a pool of names held in common by the speaker and the receiver, but two people need more than a common vocabulary in order to communicate with each other. It is also necessary that the receiver be able to use the information given by the speaker to direct his or her own attention to a likely environmental referent. The receiver must then be able to abstract information about that referent for use in conjunction with the information in the speaker's comment. For example, Tom may ask Harry, "What's that pink thing?" In order for Harry to answer, most of Tom's words, especially "pink" must be in Harry's vocabulary. In addition, Harry must be able to scan the environment for something pink and, when a possibility is located, must be able mentally and verbally to code information about it. Harry's response might be, "That's a jellyfish." Harry has then conveyed information to Tom about a referent which Harry was able to single out because Harry and Tom share a common vocabulary and because Harry possesses the attentional skill necessary to locate the quest object and to extract from it the requested information.

[1]This research was supported by National Institutes of Health grants HD-0616 and RR-00165.

[2]The experiments described here are reported in greater detail in Chapter 10 of Language Learning by a Chimpanzee: The Lana Project, D.M. Rumbaugh (Ed.). This talk was a condensation of that chapter.

[3]Now at the Wisconsin Regional Primate Research Center, University of Wisconsin, Madison, WI. 53706.

Clearly the ability to attach names to objects is fundamental to using language. When this phase of the Lana project
began in the early summer of 1974, Lana had acquired at least
rudimentary naming skills through her training in a computer-
controlled language-training situation (Rumbaugh, 1977).

Impressed by Lana's apparent grasp of the function of
names, we decided at this point to explore her ability to use
names more fully. To do so, we planned to present her with a
task which would require her to make a visual search of objects and to answer specific questions similar to the Tom and
Harry example mentioned above. We considered this a sophisticated task, and we recognized that we would first have to
demonstrate the existence of the requisite common vocabulary
before attempting to request information about a spatially removed referent via a linguistic code. The following set of
experiments was designed to test for the existence of such
skills. At the time, Lana had been in the language-training
situation for only 20 months.

METHOD

Subject

Lana, a female chimpanzee (Pan), was the only subject.
She was 3 3/4 years old at the time testing began. Her previous training included extensive practice in giving the name
or color of an object physically present or projected from a
35 mm slide. For the duration of testing (approximately two
months) Lana's food intake was limited to the rewards received
during training and testing with occasional supplements to meet
her dietary needs.

Apparatus

The computer-controlled language-training facility has
been described elsewhere (Rumbaugh, von Glasersfeld, Warner,
Pisani, Gill, Brown & Bell, 1973). Briefly, this system en-
ables Lana to control multiple aspects of each 24 hour day via
a keyboard consisting of individual keys embossed with geomet-
ric patterns. Each key represents a concept. At the time when
the experiments described here took place, the summer of 1974,
there were 75 keys on the keyboard. Unless otherwise noted, a
cardboard screen hung on the door to Lana's room during testing
periods and precluded visual contact between the subject and
experimenter.

Experiment I: Color- versus object-naming of projected
photographs. As outlined above, Lana had already learned to

name various objects and to give their colors. The first experiment addressed the question of whether she could view a projected slide of a familiar object and correctly respond to the questions, ? What name-of this or ? What color of this.

The stimuli were 36 slides which were projected directly above Lana's keyboard. Each slide was a familiar photograph of one of six possible objects on a white background, each of which had been spray painted one of six different colors. The six objects were: ball, bowl, box, can, cup, shoe. The six colors were: black, blue, orange, purple, red, yellow.

At the beginning of each trial, the experimenter would depress keys on the experimenter's keyboard to ask Lana either, ? What name-of this or ? What color of this. As the key for this was depressed, a slide was projected. To answer the question, Lana first depressed the "PERIOD" key to erase the experimenter's question and then depressed the keys corresponding to her answer. If her answer was correct, the experimenter depressed the Yes key, and a reward (generally milk, juice or a small piece of fruit) was delivered.

Over the course of four 18-trial sessions, Lana performed this task with about 80% accuracy. A generous way to estimate what percentage of correct responses would be expected by chance alone would be to assume that Lana would always respond to the correct question using the correct sentence structure (either Color of this --- or Name of this ---), but that she might guess as to which color or object key should be depressed. Since there would then be six possible keys to choose from when answering either question, the performance expected by chance alone would be about 16.7% correct. For either question type, Lana's performance was significantly above this figure ($p < .001$, exact binomial).

Experiment II: Color naming of novel objects. Experiment I demonstrated that Lana could reliably give appropriate color responses when shown slides of familiar objects. Certainly we hoped, but could not be sure, that Lana's responses stemmed from a concept of color and not simply from a knowledge of which of the six "color" responses should be paired with which of the familiar objects. If Lana indeed had a concept of color, we reasoned, then she should be able to apply her color terms to describe unfamiliar objects.

We presented Lana with 36 novel junk objects between 1 1/2 and 6 inches long in any given direction which were spray painted, if necessary, to conform to one of the six colors used in Experiment I. Color naming of novel objects was a very easy task for Lana; her performance was well above chance ($p < .001$, normal approximation to the binomial), with a mean percentage of correct responses of 87.

Experiment III: Color naming of requested object when

several objects are present. Up to this point, only one object
had been present whenever Lana had been asked to give color or
name responses. A more demanding task would involve asking
the same questions with several objects present. Then, in
order to answer correctly, Lana would have to do the following:

1. Read the question to determine what information was
being requested about what object.
2. Use the information extracted from the question to
guide a visual search and isolate the desired object.
3. Examine the reference object to determine the requested information.
4. Code the requested information into the correct
key presses.
5. Depress the necessary keys to form the appropriate
response.

For this task the actual objects photographed for the
slides in Experiment I were presented three at a time to Lana
in five blocks of 36 trials each, with 18 trials given per
session. Thus a given trial might begin with presentation of
a blue box, a yellow bowl, and a red shoe. Lana was asked to
give the color of a specific object present on that trial, for
example, ? What color of this box.   Testing was done with the
door to Lana's room open so that Lana could come over to the
door area and inspect the objects. With three objects, Lana
averaged 79% correct (p < .001, exact binomial).

The number of objects present on each trial was next
increased from 3 to 6. Very interestingly, Lana's performance
improved when the number of objects was increased, she averaged
89% correct. This improvement in performance coincided with a
behavioral change on Lana's part. Once the experimenter had
typed out the question and Lana had inspected the objects, she
went to her keyboard and typed out the first part of her answer (e.g., Color of this shoe) as before, but when six objects
were present, she then turned toward the objects once again
before finishing the sentence. This second "inspection" of
the objects was presumably a visual check which she imposed on
herself when the task of remembering the correct object without
a second look became more difficult.

Experiment IV: Naming of object requested by color when
several objects are present.   Experiment III demonstrated that
Lana could give the color of a requested object when more than
one object was present. We next asked whether Lana could give
the name of an object identified only by its color when more
than one object was present. The question posed to Lana was,
? What name-of this that's ----, the blank being filled in by
the color of one of the objects present on that trial. Two
sessions of 18 trials with three objects present were given,
followed by two sessions of 18 trials with six objects present.

This allowed each object in each different color to be the requested object once when three objects and once when six objects were present per trial.

Not surprisingly, Lana also did well on this task. Although there were twice as many chances to make errors (giving the incorrect color or name) as there were questions, Lana's over-all percentage of correct responses when three objects were present was about 92% and, when six objects were present, was about 94% ($p < .001$, exact binomial). Once again Lana's performance was not hindered when the task was made more difficult by increasing the number of objects present during each trial.

Experiment V: Intermingled color- and object-naming questions of a requested object with six objects present. We next wondered how Lana would perform if the questions ? What color of this and ? What name-of this were intermixed. A high level of performance on such a task would indicate that she could both differentiate between what was requested and answer a specific question when a number of objects were present.

In this experiment, the 36 objects were again presented to Lana in groups of 6 (one of each object with each color present) with either the question of Experiment II (? What color of this ----) or of Experiment III (? What name-of this that's ----) being asked. The question to be asked was selected randomly with the restrictions that not more than three questions of one type should be asked in a row and that, within the first 36 trials, each question should be asked 18 times. A second series of 36 trials was then given using the same stimulus arrays, but with the opposite questions being asked.

Lana demonstrated the same high level of performance when the questions were intermingled as when a series of each was asked separately (her over-all average was about 92% correct; $p < .001$, exact binomial). Clearly she both noted what information was being requested and responded correctly even when several objects were present.

GENERAL DISCUSSION

Lana's success on these tasks indicates that she can mentally manipulate abstract concepts which have been defined by means of an arbitrary code. Such manipulation is necessary if one is to scan a set of objects and mentally select one of the objects on the basis of a linguistically expressed criterion. Answering ? What name-of this that's blue when several objects are present demands more than a vocabulary containing a particular set of words. A goal-directed visual search must be initiated which is based on information deduced from the

abstractly coded question. The result of this search must
then be linguistically coded and expressed.

This would seem an impressive task for an only-very-
recently-linguistically-trained chimp, but we have no reason
to believe that it was a particularly difficult one for Lana.
Although her performance was generally less than 100% correct,
she was always well above the figure which could be expected
by guessing. Moreover, we would speculate that most of Lana's
errors stemmed from a wavering devotion to the task at hand
since, typically, the first few trials of a session were error
free.

Formal testing such as that reported here appeared arduous for Lana. At times she whined or asked to be taken outside. Certainly the sessions were useful even though it is
doubtful that Lana learned very much from them: her percentage
of correct responses was fairly constant across the sessions
comprising any given experiment. From such experiments we
gained certainty about what we suspected or hoped Lana could
do. We now know that Lana can describe a given object in more
than one way and can select her mode of description according
to the question posed; she can mentally transpose a linguistically phrased question into what is necessary for a visual
search, and, once the requested information is obtained, she
can give it via a linguistic code.

REFERENCES

Rumbaugh, D. M. Language Learning by a Chimpanzee: The Lana
     Project. New York: Academic Press, 1977.
Rumbaugh, D. M., von Glasersfeld, E.C., Warner, H., Pisani,
     P., Gill, T. V., Brown, J. V., and Bell, C. L.
     A computer-controlled language training system for in-
     vestigating the language skills of young apes. Behavior
     Research Methods and Instrumentation, 5, 385-392, 1973.

APES, ANTHROPOLOGISTS AND LANGUAGE

Gordon W. Hewes

University of Colorado
Department of Anthropology
Boulder, Colorado

I have based my title on Eric Linden's book, Apes, Men and Language, to emphasize the point that anthropologists have or should have done some serious rethinking about language and culture as a result of the primatological research of the last two decades, and especially as a result of language studies with chimpanzees and other anthropoid apes. C. A. Bramblett in a recent review article on "Ethology and Primates: some new directions for the 1970's" (1976:593-607) provides a good survey of this impact. To be sure, anthropology, at least since the mid-19th century, has not been able to confine itself solely to Homo sapiens. As we shall see, one reason for this, since the 1920's has been anthropological awareness, if not deep involvement with, the research undertaken by Robert W. Yerkes.

Speculation that apes might be somehow able to learn human language, if not speech, then by "signs", goes back a long time too. Although Rene Descartes denied reason to animals on the ground that they could never speak (or otherwise employ language), in 1637, Samuel Pepys, the English diarist and Admiralty official, after observing a large primate - possibly a chimpanzee, though Pepys called it "a great baboon", in 1661, thought it possible that such a creature might learn to communicate "by signs". In 1748, Julien Onffroy de la Mettrie was convinced that an anthropoid ape, if properly instructed, preferably by a teacher used to working with deaf human pupils, could learn speech and thus become a "perfect little gentleman".

Eighteenth century knowledge of the great apes (as Yerkes and Yerkes (1929) made clear in their The Great Apes) was confused, and chimpanzees were lumped with orangutans; the gorilla was still unknown to science. On the basis of this ignorance, Lord Monboddo, a Scottish Judge and writer on language origins (1773) supposed that apes could speak, or perhaps had simply forgotten how to speak. In 1699, the

English anatomist Tyson, after careful dissection of the brain
and voice-box of the chimpanzee, expressed puzzlement over why
such an animal could not talk.   Later anatomists such as
Pieter Camper (1779) and George Cuvier satisfied themselves
that anthropoid apes were structurally incapable of speech. In
any case, several chimpanzees and orangutans reached Europe
alive (usually only to succumb soon after their arrival) and
in almost every case, aroused speculation about their possible
language capabilities.   Buffon's two-year old chimpanzee,
which survived for a time in Paris, was probably seen by de la
Mettrie, since it was described by Buffon in 1740.   George J.
Romanes, a late 19th-century psychologist, deeply interested
in Darwinian evolution, wrote extensively on the "evolution of
the mind" in men and animals, and mentioned that a chimpanzee
at the London Zoo had apparently understood a good deal of
what her keeper said.   Romanes was subsequently half-forgotten
by later comparative psychologists, not only because he was
not a rigorous laboratory experimentalist, but because he
utilized anecdotal material to exemplify his arguments about
mental evolution in animals.   In the light of what is now
known about the language accomplishments of Washoe, Sarah,
Lana, Koko, et al., his work makes very interesting late 20th
century reading.

     The American pioneer student of ape and monkey "Langu-
age" as he called it, was Richard Lynch Garner, a quixotic
figure, almost pathetic in his conviction that he was decoding
the speech of his primate friends.   He went in the 1890's to
French West Africa, equipped with the latest Edison cylinder
phonographic recording instrument, and a large iron cage to be
set up in the rain-forest from which vantage point he could
watch the apes and monkeys visiting him.   Garner's work is
full of simple anthropomorphizing about animal behavior, and
outpourings of sentimentality, but R. M. Yerkes recognized
some important leads in Garner's books nevertheless.

     Yerkes tells us in his <u>Chimpanzee: a laboratory colony</u>,
(1943), that the general notion of a laboratory-field station
for comparative psychobiology came to him as early as 1900
when he was still a Harvard graduate student.   However, by the
time his textbook, <u>Introduction to Psychology</u> appeared in 1911,
apes and language seem to have been far from his mind.   Indeed
he refers to language in just one sentence (p.141), to wit:
"Doubtless the psychological significance of the way in which
language develops is of great importance", which he footnotes
to a then recent translation of Herman Ebbinghaus' <u>Abriss der</u>
<u>Psychologie</u> (Leipzig, 1908; English version, translated by
Max Meyer, New York, 1911).   Ebbinghaus devoted a long section
to language, which is termed a "complication of mental life".
Curiously, neither the young R. M. Yerkes nor Ebbinghaus
mentions the inordinate interest in the phenomena of language

on the part of the acknowledged dean of experimental psychology, the great Wilhelm Wundt, whose writings after 1900 or so
were turning more and more toward the fields of linguistics
and anthropology.

In 1914, Yerkes began corresponding with Wolfgang Köhler,
one of the moving figures in the Gestalt school of psychology,
who by then was working at the anthropoid station on the Island
of Teneriffe in the Canaries, and where he carried out the
world-famous studies of chimpanzee intelligent behavior, a
landmark in the study of the cognitive capacities of the higher
primates. Yerkes had become more interested in the possibilities of systematic work with apes and monkeys, especially
the former, stimulated by the notion of teaching apes to speak,
a project which was actually carried out and published in 1916
by W. H. Furness (and preceded in 1913 by a project with a
gibbon by Louis Boutan). Meanwhile, Yerkes learned of the
birth of a chimpanzee (one of the first cases, if not the
first, in captivity) at the private estate of Senora de Abreu,
Quinta Palatino, outside of Havana, Cuba in 1915.

Shortly after the first World War, Köhler's work on the
mentality of apes (issued in Germany in the unpropitious year
of 1918) became available, along with reports from Nadia Kohts
in Moscow of her work with the young chimpanzee Joni (1923),
and Yerkes acquired two young chimpanzees with which he worked
for a time in New Hampshire, as well as in Washington, D. C.
(also 1923). He also managed to visit Havana and was deeply
impressed by the possibilities of a large scale chimpanzee
laboratory colony. Although Robert Yerkes was interested in
many aspects of anthropoid ape behavior as a suitable field
for comparative psychology, his work indicates a persisting
interest in language or language-related aspects. In his work
which is mainly a description of Senora de Abreu's ape colony,
Almost Human (1925) is an entire chapter on what he calls
"voice and language". In the same year he and Blanche W.
Learned's Chimpanzee intelligence and its vocal expressions
appeared, which focuses on "anthropoid speech" (by now he was
carefully distinguishing these two terms), in which Learned
provided detailed protocols of chimpanzee vocal calls, using
standard musical notation along with vowels and consonants of
the Latin alphabet. In his 1943 work on the Yale chimpanzee
colony, then at Orange Park, Florida, which had just been reorganized (1942) under Yale, Harvard and foundation sponsor-
ship, Yerkes devotes a chapter on what was now called, with
more fashionable restraint, "language and symbolism".

R. M. Yerkes tried, quite unsuccessfully, to replicate
at least W. H. Furness' speech-training, which had involved
principally a young orangutan, with the two young chimpanzees,
Chim and Panzee (1925:175). Despite this failure, he nevertheless wrote (ibid.,p.180):

"I am inclined to conclude from the various evidences
that the great apes have plenty to talk about, but no
gift for the use of sounds to represent individual,
as contrasted with racial, feelings or ideas. Perhaps
they can be taught to use their fingers, somewhat as
does the deaf and dumb person, and thus helped to acquire a simple, nonvocal 'sign language'."

This suggestion was of course, as we all know, precisely
on the right track, even if all of the experimental languages
taught to apes thus far are not similar to sign languages of
the deaf; they do involve using the fingers, in manipulating
tokens, or in pushing buttons on a computer console, where
they do not employ sign-language as such.

Just at this time, but doubtless in a field then far
distant conceptually from Yerkes and his colleagues, another
gap was being closed between man and apes, although nearly
twenty-five years would pass before the mounting evidence
would begin to force changes in anthropological thinking.
Raymond Dart, in South Africa, described the first specimen of
Australopithecus (in 1925), and major accession to the later
hominid fossil record began to come to light in northern China
(1927 - 1937). The illustrations which had been fostered by
the Piltdown fraud were to die slowly among physical anthropologists and anatomists, but in the long rum (by the early
1950's!) the combined impact of the australopithecine evidence
now augmented by further finds from near Johannesburg, and
Weidenreich's meticulous studies of Peking Man were to under-
mine the fundamentally non-Darwinian hope that the fossil
record would support an almost total separation of the ancestry of man from that of the surviving Pongidae, and that an
almost miraculously modern kind of human brain had appeared
very early in the paleontological record (at least, quite
early in the Cenozoic).

Meanwhile, after the founding of 1930 of the Orange
Park laboratory station, and under Yerkes' direction, an
important series of ingenious experiments were undertaken with
chimpanzee subjects, on the effects of surrogate rewards (the
famous poker-chip rewards to be used in a so-called Chimpomat),
and on a search for a symbolizing function which could reasonably be supposed to underlie human language capacity, in which
such distinguished investigators as H. W. Nissen, A. H. Riesen,
and M. P. Crawford, participated. Meredith P. Crawford, whose
experiment involved cooperative problem-solving by a pair of
chimpanzees who had to haul a heavy food box toward their cage
by pulling on separate ropes, noted in 1941:

"It may be that an important transitional step in the
development of language behavior lies between the
direct orientation of one animal by another through
bodily manipulations and indirect orientation through

pointing toward a distant object."

From this report, Yerkes took a set of photographs (1943: pl.42) captioned "Two chimpanzees communicating by gesture in their cooperative task." These studies came, to be sure, at a time when the attention and energies of most anthropologists were quite otherwise engaged, receiving little comment except (as I shall note) from Professor Earnest A. Hooten of Harvard.

When Yerkes retired from his directorship of the then Yale Primate Biology station at Orange Park, the reorganized institution was placed under the direction of Professor Karl Lashley, Professor of Neuropsychology at Harvard, and one of the leading behaviorists of the time. I do not mean to suggest that Lashley completely switched the research orientation of the work with the great apes at that time, but his position had been for many years, as expressed in his many writings, that while mental experiences <u>may exist</u> (my emphasis) in human beings, they are without scientific utility - that is, as phenomena capable of being investigated. In the light of what is now coming to be known about cortical localization of brain functions, and cerebral lateralization, especially in connection with speech, it is also worth recalling that Lashley was famous in psychology at that period (if not among all medical neurologists and neurosurgeons) for his insistence on the equipotentiality of the cerebral cortex. It was several years after all this, to be sure, that Keith and C. Hayes carried out their experiment at Orange Park with the chimpanzee Viki, whom they raised in their home, and whom they hoped might acquire spoken language.

It was not Garner's brand of rather credulous, unsystematic field-observation of animals, or any one other piece of work along such lines, which shifted most animal psychology into the laboratories from around 1900 on, into the harnessed salivating dogs of Pavlov, or the innumerable rats running mazes or pressing levers which characterized comparative, experimental psychology for so long, especially in the United States. Given the tasks which were studied, it is probably well that only very small numbers of expensive and increasingly scarce higher primates were not required. It was part of a Kuhnian paradigm shift, going far beyong psychology, and represented in purest form in the Logical Positivist philosophers, and which, although the lines of influence remain obscure, also deeply transformed linguistics. Mental phenomena, at least under that heading, disappeared from the textbooks, and topics like imagery and cognition virtually vanished from <u>Psychological Abstracts</u>, along with the term "instinct", though "drives" remained acceptable. In linguistics even more surprisingly, <u>meaning</u> was dispensed with, and left to what was considered almost the lunatic fringe of semantics. In cultural anthropology, from the late 1920's on,

and especially in the United States, an extreme form of
cultural relativism arose, much like linguistic relativism of
the kind known as the Sapir-Whorf hypothesis. In the study of
human non-verbal behavior, the existence or even the possibility of pan-human emotional expressions was denied (cf. Weston
LaBarre, 1947, and the writings of Ray L. Birdwhistell), just
as language universals were not considered worth looking for.
Darwin's interest, expressed in his 1872 work on Expression of
the emotions in man and animals, was quite forgotten, or if
rediscovered by a student, hastily returned to the shelf as an
embarrassingly unscientific deviation.

     Direct field study of primate behavior was carried out by
a handful of investigators, not in the main stream of their
disciplines.

     Where were anthropologists in all of this? Well out of
touch or sight of monkeys and apes, for the most part! In
1928, Alfred Louis Kroeber had published a thoughful paper on
"Subhuman Culture Beginnings", Quarterly Review of Biology,
vol. 3, which Yerkes must certainly have read when it came
out. In this paper, Kroeber reviewed the work of Wolfgang
Köhler, and suggested some applications of Köhler's (and
others) findings to proto-human Stone Age culture. This was
better than total neglect, but scarcely enough. By and large,
subsequent professional anthropological concern with nonhuman primates was chiefly limited to their skeletal remains -
bones and teeth, cranial measurements, and the like, except
for Earnest A. Hooton of Harvard. Although Yerkes had voiced
his opinion in 1925 that great apes probably had plenty to
talk about, Kroeber, after Franz Boas the doyen of American
anthropologists, thought just the opposite: they do not speak
because they have nothing to say. Ludwig Wittgenstein, around
1933, is quoted as observing that if lions could speak, we
wouldn't be able to understand them. It was probably Kroeber,
rather than Hooton, whom Yerkes had in mind in the 1943 book
on the chimpanzee colony, when he disagreed with the "anthropologists dictum" that the chimpanzees, though eminently
sociable and socializable, were cultureless. Yerkes begged to
differ, but admittedly, not from a very strong empirical
position. This was a dozen years before the Japanese primate-
watchers would discover proto-cultural transmission of learned
sweet-potato washing behavior in Japanese macaques, and even
longer before Jane Goodall would produce her photographs of
wild chimpanzees in the Gombe extracting termites from their
hills with previously decorticated twigs.

     Man, more and more anthropologists were saying, was
virtually instinct-free; if there were any human predispositions stemming from our genes, they lay wholly in such controversial behaviors as the infantile Babinski reflex, and were

not even proper instincts by the definitions of the former
super-instinctivist psychologists such as William McDougall.
The biological behavioral continuum was broken at some un-
specified time in the course of hominid evolution, since when
mankind has become practically entirely a creature of learning
or experience, a perfect tabula rasa as Hume had envisaged it
back in the 18th century. The fact that such views seemed to
contradict evolutionary biology did not seem to disturb their
advocates.

Hooton of Harvard, as indicated, did persist in paying
attention not only to the skulls and bones of primates, but
to reports of their behavior, in the field and from the
laboratory, though Hooton himself did not take a personal part
in field and laboratory primate research. Significantly,
although their contents often deal with many other subjects,
his books included Up from the Ape (1st. ed., 1931), Apes,
men and morons (1937), Why men behave like apes and vice versa
(1940), and a work dealing mainly with apes and monkeys,
Man's poor relations (1942). One can consult these volumes to
measure the impact of, among others, Yerkes and his colleagues
working at Orange Park, and elsewhere. Hooton, at least, was
not wholly taken in by the fashionable culturology of the day,
in denying the relevance, except perhaps morphologically, of
our biological kinship with non-human primates.

Just as there had been a profound paradigm shift affecting not only psychology, but biology, anthropology, and even
philosophy, along with linguistics, alienating those who persisted in wanting to study apes and monkeys, the pendulum
swung back, starting some time in the late 1960's (and perhaps
first of all in linguistics), although the mushroom growth of
primatology from around 1950 seems to have been important too.
This shift was accompanied by the discovery that, partly ob-
scured by what had happened in Germany politically and then
militarily, some Germans and Austrians had been finding out
some astonishing things about greylag geese and herring gulls,
and long before World II, for that matter, about so-called
"bee language", mediated by peculiar dances. In short, one
of the factors in the paradigm shift was the emergence of
ethology, including even human ethology (cf. the work of
Eibl-Eibesfeldt, et al.). In a recent review chapter, Robin
Fox and Usher Fleising observe that the essence of the human
ethological approach lies in acceptance of the modern syn-
thetic theory of biological evolution as the "master paradigm"
including that which seems to underlie the phenomena of tool-
using and language-using. If we can manage to clear away the
details which occlude our understanding of the broad picture,
it seems that what really was involved for about fifty years
in several sciences, including psychology, anthropology, and

linguistics, to name three, <u>was a retreat from biological evolutionary thinking</u> - never quite openly admitted, or perhaps even consciously recognized by those who could have been counted upon to join in regarding the Darwinian revolution as a major achievement in science with a capital 'S'. Cultural anthropologists and linguists, while still paying lip-service to Darwinism, had in fact concluded that evolution had become irrelevant for mankind a very long time ago, some time in the Lower Paleolithic presumably (cf. Philip Hines, in Current Anthropology <u>17</u>(3): 521).

One could not describe present-day anthropology as so detached from the implications of biological evolution, even though some of the implications may provoke certain anthropologists to mayhem or worse, nor, after the Chomskian revolution, which so strongly emphasized the innate character of man's propensity for language acquisition, (and hence, as something inborn, hardly immune from the machinations of those dastardly genes, or modifications in the DNA molecule!) linguistics.

While the bulk of the work and reportage on the primate language research has been by psychologists, we anthropologists have been taking increasing notice, and one finds substantial concern in anthropological circles, journals, and at anthropological meetings. Two of the three organizers of the 1975 New York Academy of Sciences Conference on the origins and evolution of language, in which there was extensive discussion of the ape language research, were anthropologists (Lancaster and Steklis). <u>Current Anthropology</u> has proved a hospitable journal for the airing of views pro and con, about language in apes. I am also glad to report that during the last few years, nearly all the new introductory anthropology textbooks I have examined contain significant references to the ape language experiments - sometimes extensively treated with relevant illustrations, charts, and tables, usually in sections dealing with the nature of human language, and often with reference to the long neglected topic of the origins of language.

I conclude with some possibly gratuitous suggestions for further research on language and language-related behavior with ape subjects, which I hope may develop if this extremely promising endeavor can continue to be supported and allowed to expand.

1. At the neurological level, just where is ape language being localized in the brain? In one or in both hemispheres equally? Does ape language have any tendency to become linked with preferential handedness, as it seems to in Homo sapiens? Given that the languages imparted to apes are non-vocal, do

they then exhibit similar neural patterns expectable in deaf
users of sign language?

2. Related to No. 1, above, are the developmental stages of
acquisition of language or language-like behavior the same in
young children and young apes? John Lamendella, and also
Suzanne Chevalier-Skolnikoff, both in California, have been
working in this area already, i.e., developmental anthropoid
psycholinguistics.

3. A major expansion of the use of language systems inculcated
in apes to study many aspects of their cognitive world, such
as we normally study in human subjects through verbal means.
This would be in line with the suggestions of Donald R. Griffin
in his recent article in American Scientist, 64:530-535, and
in his book (1976) The Question of Animal Awareness.
This kind of spin-off would be of great importance for psychology generally, and not just in the realm of anthropoid psycholinguistics. Chimpanzees and other apes, so far as they
remain available for research, and do not perish entirely, are
experimentally manipulate in ways not ethically permissible
with human subjects.

4. Work on the obvious topic of the transmissibility of
language to apes, to other apes (infant or adult) through more
or less normal ape-to-ape means without further human intervention. This is suggested research topic which comes up
regularly in the classroom.

5. How does possession of even rudimentary language affect or
enhance cognitive processes, and, by extension, successful
adaptation to the environment. Does acquisition of language
permit construction of more adequate mental maps of complex
terrain, for example? Does possession of language, even at a
rudimentary level, lead to greater tool-making or tool-using
capacities?

6. The language systems used heretofore in work with anthropoid apes, though not precisely replications of the structure
or morphology of English (or some other closely related Indo-
European language, such as French, which to some extent still
underlies ASL) raise the problem of the possible different
outcomes were apes taught languages derived from or patterned
on some natural human languages of radically different
structure, such as Navaho, Kwatiutl, Eskimo, or some of the
Highland languages of New Guinea. Here, again there are
research possibilities feasible ethically in non-human subjects, which we could not properly carry out on young human
children, for example.

7. Does inculcation of a visual/manual language in an ape
facilitate later acquisition or even simply receptive acquisition of human spoken language? Roger Fouts has done some work

on this topic, but it certainly could be expanded. Most of
those who have had prolonged contact with chimpanzees, gorillas,
or orangutans report a considerable degree of what seems to be
receptive understanding of human speech (whether more or less
than what most dog-owners anecdotally report remains to be
carefully determined). The failures of the earlier language-
teaching experiments of Furness, the Hayes, and of R.M. Yerkes,
for example, were with apes without opportunity to learn a
previous non-vocally-based language.

8.   Related to No. 7 above, it should be possible to try to
teach apes to read and write - not necessarily using handwriting! Blissymbolics, a system of highly iconic signs now
being successfully used for cerebral palsied patients and with
some mental retardates, might be used rather than words in the
Latin alphabet, or in Chinese characters. There are time-
binding aspects of the use of writing which might have other
interesting effects if apes could be taught to use such systems. Not least would be the possibility of imparting a much
richer store of knowledge about the external world to apes
through both the visual and written media, obviously with
potential spin-offs for problems affecting human education,
particularly for retarded individuals.

        Most of these suggestions do not have much promise of
jobs in ape communication for anthropologists. However, in
some such projects, the occasional presence or consultation
of an anthropologist might be useful, chiefly to get away from
culture-bound notions, or from the more limited time-perspec-
tives which anthropologists suppose characterize most of their
non-anthropological professional colleagues.

        For anthropology, there is great hope that as this
research with apes goes on and expands, we may be helped to
recover the holistic approach which once characterized our
discipline, notably in the United States (it never was very
holistic on the European Continent), grounded in a richly
informed evolutionary perspective.

REFERENCES

Bramblett, C. A., 1976, Ethology and Primates: Some new
        directions for the 1970's. American Anthropologist 78:
        593-607.
Fox, Robin and U. Fleising, 1976, Human Ethology. Annual Re-
        view of Anthropology, 5: 265-288.
Yerkes, Robert M. 1925, Almost Human, New York, The Century
        Company.
Yerkes, Robert M. 1943, Chimpanzees, a laboratory colony.
        New Haven, Yale University Press.

Yerkes, Robert M. and Learned, B. W.   1925. Chimpanzee
     intelligence and its vocal expressions.   Baltimore,
     The Williams and Wilkins Company.

INTRODUCTION: CHIMPANZEES AS BIOMEDICAL MODELS

L. D. Byrd

Yerkes Regional Primate Research Center
Emory University
Atlanta, Georgia

Good morning, and welcome to the Third Session of the Yerkes Centennial Conference, entitled "Chimpanzees as Biomedical Models". In this session, we hope to identify several important ways in which studies in the chimpanzee have contributed and can continue to contribute to a better understanding of health-related problems in animals and humans.

When one considers the numerous areas of research in which the chimpanzee has been found to be of special value for studying biomedical problems, the task of identifying a few representative areas becomes difficult. We know from the observations and experiments conducted by many of our honored guests in the Yerkes Laboratories at Yale and Orange Park that the chimpanzee is an excellent model for studying reproductive biology, drug addiction and drug dependence, cardiovascular function and circulatory phenomena, hormonal correlates of behavior, hepatic dysfunction in cross-circulation, immunological responsiveness and visual and auditory function, to mention a few.

In view of the foregoing, the areas of investigation represented by the speakers this morning cannot be and are not intended to be exhaustive of the various models the chimpanzee can offer to biomedical research, and I hope that no one present feels that his or her special area of interest is neglected. I think that the five areas to be highlighted in this session can be regarded as representative of those in which the chimpanzee has become a valuable subject model in the biomedical field.

Supported by U. S. Public Health Service Grants DA 01161 and RR 00165, Division of Research Resources, National Institutes of Health.

Before introducing the first speaker, I wish to take a few minutes to recall for the audience the kind of special requirements that can arise when one is using large primates, and especially the great apes, as experimental subjects. Many of us were reminded of the special considerations associated with conducting studies in conscious apes when Drs. Elder and Spragg described some of their classic studies in the chimpanzee yesterday morning (Elder, 1977; Spragg, 1977). You will recall that the basic approach was to train the animals to cooperate and put themselves in a position that permitted the conduct of experiments.

For a number of years now, I have been interested in the way pharmacological agents or drugs can act on the central nervous system and, consequently, alter behavior. The behavior of interest is a conditioned motor response whose frequency of occurrence is determined by past consequences, i.e. conditioned operant behavior. The application of conditioned behavior to studies of the behavioral effects of drugs dates back several decades. In 1937, Skinner and Heron published a paper describing the effect a commonly used substance, caffeine, had on conditioned behavior in the rat. Compared to the total behavioral output obtained in a group of rats when saline was injected, the number of responses during a one-hour period nearly doubled when ten milligrams of caffeine were injected. The Skinner and Heron paper represents one of the earliest experiments in behavioral pharmacology using conditioned operant behavior.

Scientists have accumulated substantial information about the behavioral effects of a variety of drugs in a diversity of species since Skinner and Heron published their paper in 1937. The big impetus came, of course, during the early 1950's when chlorpromazine was observed to have beneficial effects as a tranquilizer in troubled humans. The resulting surge in research activity produced increased information about the behavioral effects of drugs in mice, rats, birds, and several species of monkeys. However, few experiments were conducted in the chimpanzee and, consequently, little is known about the effects of drugs in this primate. Inspection of Rohles' Topical Bibliography of the Chimpanzee reveals only six references under the heading "Pharmacology" as recently as 1962. The paucity of information available on the chimpanzee prompted me to begin studying the behavioral effects of drugs in this ape and to determine the generality of the effects observed in other nonhuman primates.

One has to work with great apes to appreciate the kinds of problems associated with these large animals. For example, it is relatively easy to pick up a small rodent, cat or small monkey and inject a drug without traumatizing or getting injured by the animal. Treating a great ape in a similar manner

is simply not feasible. A drug can undoubtedly be injected if
the chimpanzee is physically restrained in a squeeze cage, but
this experience is usually disruptive and can confound the
effects of the drug one is interested in studying. It seems
more appropriate and productive to adopt a policy similar to
that described by Drs. Elder and Spragg and train the chimpanzees to cooperate voluntarily and engage in those behaviors
conducive to the execution of the study. The classical work of
Dr. Spragg in developing morphine addiction in the chimpanzee
provided evidence that the chimpanzee would readily tolerate
repeated injections of an addictive substance without physical
restraint (Spragg, 1940). However, the experiments of interest
to me involved non-addictive drugs or an injection schedule
that avoided the development of addiction or physical dependence. Fortunately, Dr. Charles Ferster's success in training
an adult chimpanzee at the Yerkes Laboratories to tolerate a
fingerprick to obtain blood for cell counts provided additional
evidence that it was feasible to consider training chimpanzees
to accept intramuscular injections of inert substances and of
drugs not known to be addictive.

Our efforts in training chimpanzees to extend an arm and
accept an intramuscular injection were successful. The behavior of extending the arm and accepting the injection was reinforced with a piece of food. After 2-3 weeks of intensive
training, the chimpanzees readily permitted the injections on
a regular schedule. The extension of the arm prior to each
experimental session became as routine as any other aspect of
the laboratory procedure and the animals cooperated readily
without any indication of discomfort or distress. Consequently,
the changes in conditioned behavior observed during a session
following the injection of a drug were more convincingly
attributed to the drug than to some aspect of the procedure of
getting the drug into the animal.

A similar procedure was adopted for obtaining daily
weights of chimpanzees. A platform scale large enough to weigh
an animal while he sat in a cage was not available when I began
these studies and, therefore, an alternative procedure had to
be developed. A general purpose scale of the type frequently
used in produce and meat markets was suspended above the animal's living area. The chimpanzee was trained to grip a piece
of rope attached to the scale and hang motionless long enough
for the observer to read the weight on the scale. This technique allowed us to obtain a precise measurement of the animal's
weight, and then use the weight to determine the exact amount
of drugs to inject.

Some of the drugs studied in the chimpanzee have yielded
interesting and unexpected results. Chlorpromazine, a tranquilizer used extensively in treating aberrant behavior in
humans, had effects qualitatively different from those reported

for monkeys. Whereas chlorpromazine typically decreases responding in monkeys (Kelleher and Morse, 1968), the drug increased responding under similar procedures in the chimpanzee (Byrd, 1974). Morphine is another drug that characteristically decreases responding in monkeys (Byrd, 1976; Goldberg, Morse and Goldberg, 1976; Woods and Schuster, 1971), yet markedly increased responding in chimpanzees (Byrd, 1975). The basis of these contrasting effects is unknown and subject to speculation. One might presume that they are related to differences in metabolic disposition or rate of disposition or differences in active metabolites, since other drugs, e.g. d-amphetamine (Byrd, 1973), do not have dissimilar effects in monkeys and apes.

My interest in making these brief comments is to call attention to several considerations. One is to point out the way in which work done at the Yerkes Laboratories years ago has provided a foundation for contemporary research. We will see the contributions from earlier investigators become more evident as the following five speakers address this audience. Secondly, I want to emphasize how little we know about some aspects of the chimpanzee, the most widely used great ape. If we are to entertain the notion that the chimpanzee can be an excellent model for studies of behavior vis-a-vis human behavior, we must know more about this animal. Certainly, little is known about the behavioral effects of drugs. The usefulness of the chimpanzee as a biomedical model will be limited by the extent of our knowledge of the chimpanzee. Thirdly, I hope my comments have indicated the type of interesting effects observed in our studies of the chimpanzee and the way some contrast with effects in other nonhuman primates.

To obtain several perspectives on the value of the chimpanzee as a model for biomedical research, let us move now to the five speakers of the Third Session.

REFERENCES

Byrd, L. D.: Effects of d-amphetamine on schedule-controlled key pressing and drinking in the chimpanzee. J. Pharmacol. Exp. Ther. 185: 633-641, 1973.

Byrd, L. D.: Modification of the effects of chlorpromazine on behavior in the chimpanzee. J. Pharmacol. Exp. Ther. 189: 24-32, 1974.

Byrd, L. D.: Contrasting effects of morphine on schedule-controlled behavior in the chimpanzee and baboon. J. Pharmacol. Exp. Ther. 193: 861-869, 1975.

Byrd, L. D.: Effects of morphine alone and in combination with naloxone or d-amphetamine on shock-maintained behavior in the squirrel monkey. Psychopharmacology 49: 225-234 1976.

Liuer, J. H.; Robert M. Yerkes and memories of early days in the laboratories. In: Progress in Ape Research, G. H. Bourne (ed.) (Proc. Symp. Yerkes Centennial Conf., Oct. 1976, Atlanta, Ga.). Academic Press: New York, 1977.

Goldberg, S. R., Morse, W. H. and Goldberg, D. M.: Some behavioral effects of morphine, naloxone and nalorphine in the squirrel monkey and the pigeon. J. Pharmacol. Exp. Ther. 196: 625-636, 1976.

Kelleher, R. T. and Morse, W. H.: Determinants of the specificity of behavioral effects of drugs. Ergeb. Physiol. 60: 1-56, 1968.

Rohles, F. H., Jr.: The Chimpanzee. A Total Bibliography, Technical Documentary Report No. ARL-TDR-62-9. 6571st Aeromedical Research Laboratory: Holloman Air Force Base, New Mexico, 1962, 312 pp.

Skinner, B. F. and Heron, W. T.: Effects of caffeine and benzedrine upon conditioning and extinction. Psychol. Rec. 1: 340-346, 1937.

Spragg, S. D. S.: Morphine addiction in chimpanzees. Comp. Psychol. Monogr. 15: 1-132, 1940.

Spragg, S. D. S.: Reminiscences of early days in New Haven and Orange Park. In: Progress in Ape Research. G. H. Bourne (ed.). (Proc. Symp. Yerkes Centennial Conf.), Oct. 1976, Atlanta, Ga.). Academic Press, New York, 1977.

Woods, J. H. and Schuster, C. R.: Opiates as reinforcing stimuli. In: Stimulus Properties of Drugs, T. Thompson and R. Pickens (eds.). Appleton-Century-Crofts: New York, 1971, 163-175.

ACUTE EFFECTS OF STIMULANTS AND DEPRESSANTS ON SEQUENTIAL LEARNING IN GREAT APES

Walter A. Pieper

Department of Psychology, Georgia State University
and Yerkes Regional Primate Research Center
Emory University, Atlanta, Georgia

The primary purpose of the research program described in this paper was to develop behavioral methodology which could predict the potential behavioral toxicity in nonhuman primates. For the purpose of this discussion, behavioral toxicity is operationally defined as a reduction in acquisition performance which is measured by a sequential learning task.

In working with large nonhuman primates, the sheer physical size of the subjects places several constraints on the types of methods used to test their learning capacities. Since they are larger and more difficult to house and maintain in relatively large numbers, it is imperative that the maximum amount of data be obtained from each subject and this, of necessity, means using each subject as its own control. Measuring repeated learning in individual subjects requires that the task being used to measure such learning be designed so that acquisition can be repeated many times in a highly controlled situation.

The sequential response task used in the present program was first described by Boren and Devine (1968). It has subsequently been modified for use with great ape subjects, and

This research was supported by Drug Enforcement Administration Contract No. J69-10. Basic support of the Yerkes Regional Primate Research Center was provided by PHS Grant RR 00165 of the Division of Research Resources, National Institutes of Health.

is uniquely suited to measuring within-subject repeated acquisition in three primary respects. First, each subject is required to learn a relatively complex response sequence rather
than a simple visual discrimination. Second, stimulus fading
procedures are used to guide acquisition of the sequential
response chain in a programmed manner. Initially, cues related to the brightness of the stimulus lights above the levers
are used to shape the animal's behavior. However, as the subject responds correctly, the light cue is gradually faded and
the length of the chain is extended. Third, and of critical
importance, is the fact that while the basic nature of the task
is the same from session to session, the subject is required
to learn a new lever sequence on each test day. In this way,
the learning deficit or enhancement associated with the intake
of various psychoactive drugs can be assessed in individual
subjects.

METHOD

Sequential Learning Task

A Lab-Care Co. cage (Model 5022), 1.5 m high, 1.0 m wide
and 1.4 m long, was modified by replacing a section of one side
with a 6 mm aluminum alloy plate (29 cm high x 46 cm long)
which served as a mounting panel for six Lindsley response
levers (Model 6310, Ralph Gerbrands Co.) mounted in a horizontal array on 7.5 cm centers. A stimulus light with a diameter
of 6.2 cm was mounted 10 cm directly above each lever.
These lights served as visual cues signalling for the
animal the correct lever. The saliency of these brightness
cues was varied across eight levels of illumination, ranging
from high contrast (the correct lever fully lighted with the
other five off) to no contrast (all levers, including the
correct one, at full brightness). Centered directly below the
levers was a food receptacle into which M&M candy or peanut
reinforcers were delivered from a dispenser. These reinforcers
were dispensed each time the animal pulled the final lever in
a correct sequence of levers. When an error was made, a seven
(7) second timeout went into effect during which a buzzer was
sounded and the house lights and all lever lights were extin-
guished. Testing began each day with maximum light cues
followed by a gradual fading out of the discriminable brightness
feature, thereby forcing the subject to learn the position of
the correct levers. The length of the lever pulling sequence
was also systematically increased from one lever to six levers.
The fading procedure was first used with the shortest sequence
length of one. Each correct lever response was followed by a

reduction in the brightness cue by one level while each error
resulted in the light cue "backing up" to the level used on the
previous trial. At the eighth level, all stimulus lights were
identically illuminated and, consequently, no longer functioned
as a cue. Therefore, a correct response at the eighth level
indicated that the animal had learned the correct position of
the lever(s). Thus, the brightness cue was systematically
faded out through eight equal intensity changes, so that the
stimulus lights provided no cue to the correct sequence of
levers when acquisition of the behavioral chain was completed.
When the subject correctly completed sequence one, the sequence
was increased to a two-lever chain. The subject was now required to pull two levers in the correct sequence to obtain
reinforcement. Again, maximal brightness cues were used on the
initial trials but were progressively faded out with each correct response. When the subject successfully completed the
sequence of two levers at the eighth level of illumination, demonstrating that he had learned to discriminate the correct
two-lever sequence by position, the sequence was lengthened to
three levers and the process repeated.

Daily test sessions were terminated either when 50
minutes had elapsed or when the subject had successfully completed a sequence of six levers at the eighth light level,
whichever occurred first. Since a new lever sequence was presented each day, a new sequence was learned each day. Thus,
this task provided learning data for each day of testing.
Therefore, learning (acquisition) on those days when a particular drug was administered could be compared with intervening
control days when only placebo was given.

Multiple dependent measures can be derived from this
task. Chain length completed, interresponse time and optimum
divided by actual number of responses are included in this report. Additional measures, including number of reinforcements
obtained, errors at each light level and errors at each chain
length, are presently being evaluated for possible future analysis.

Subjects Tested

A total of eight adult or young adult great apes have
been tested on this task: two female and four male chimpanzees
(Pan troglodytes) and two male orangutans (Pongo pygmaeus). At
the present time, six of these animals (4 male, 2 female) are
serving as subjects. Their estimated ages vary from 10-30
years (average=18 years) and their current weights range from
30-85 kilograms (average=53.3 kg).

Drugs and Procedures

Data on eight drugs in the general class of depressants or tranquilizers and ten compounds generally classified as stimulant or anorectic drugs (Table 1) are included in this report and will subsequently be referred to simply as "depressants" or "stimulants". Typically, three subjects are tested on low,

| Name of Drug | Percent of Control on High Drug Dose Days | | |
	Maximum Chain Length	Interresponse Time	Response Efficiency
Butabarbital (4)	87	100	80
Diazepam (3)	58	214	44
Glutethimide (2)	95	144	73
Halazepam (3)	96	119	62
Meprobamate (2)	97	154	82
Phenobarbital (2)	87	131	75
Scopolamine (2)	105	73	69
Secobarbital (3)	93	112	84
Benzphetamine (4)	40	630	141
Chlorphentermine (3)	76	237	156
Clortermine (3)	92	141	150
d-Amphetamine (3)	61	336	118
Diethylpropion (4)	46	267	125
Mazindol (3)	52	2184	93
Methamphetamine (3)	84	238	124
Methylphenidate (3)	39	93	95
Phendimetrazine (3)	67	196	124
Phentermine (3)	62	125	208

Table 1. Effects of stimulant and depressant drugs at the high dose on maximum chain length, interresponse time and response efficiency. Number in parentheses following drug drug indicates number of subjects included in analysis.

medium and high doses of each drug. A minimum of two control
days (vehicle only) intervenes between drug days. Each ape is
tested in at least three sessions on each dose of the drug.
Determination of the low doses is based on the clinical literature and is usually slightly higher than the recommended clinical dose for humans. When information on clinical doses is
unavailable, the low dose is estimated from similar drugs.
The high drug doses are determined behaviorally and physiologically and are defined as the highest dose short of severe
behavioral or physiological toxicity. That is, the dose is
increased until the animal begins to show marked signs of
physiological toxicity (e.g., vomiting, hyperactivity, sleep,
etc.) or behavioral toxicity (e.g., failure to respond). The
amount of drug administered is then lowered slightly and defined as a high dose. The medium dose is approximately equidi-
stant between the low and high doses.

CHAIN LENGTH   RESULTS

      Percent change from control performance on the chain
length measure following the high doses of drug is presented in
Figure 1. There appears to be a marked difference between
stimulant and depressant drugs on this measure. While animals
receiving depressant drugs show relatively minor decreases
from control levels (less than 15% in most cases), most stimulant drugs produced large decreases in maximum sequence length
completed (over 35% in half the apes tested). The control days
used in this and subsequent analyses are those days immediately
preceding a drug administration day. Although a minimum of two
control days are included between drug doses, additional control days are scheduled if the subject does not achieve criterion
baseline performance on the second control day.

INTERRESPONSE TIME

      In Figure 2, the mean interresponse times (IRT) are de-
picted as percent changes from control levels. With few excep-
tions, administration of these high doses of drug was associated with increases in mean IRT; however, this change is much
more pronounced with stimulant drugs than with depressant compounds. Increases in this dependent measure probably reflect
increased pausing by the subjects rather than actual changes
in rate of responding. Visual observations of subjects during
testing reveal that subjects tend to stop working on the lever
task and move to another area of the testing cage more frequently if they have received a stimulant drug prior to testing.
This is much less likely to occur if the animal has received a
high dose of a depressant drug.

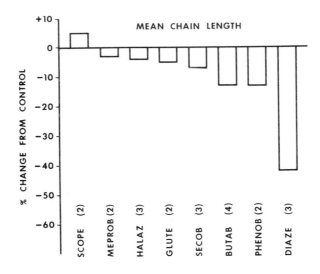

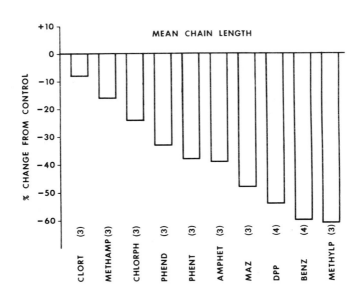

Fig. 1.   Average chain length achieved on high dose drug days expressed as percent change from control. The number of subjects included in each bar is indicated within the parentheses next to the drug name (complete names in Table 1).

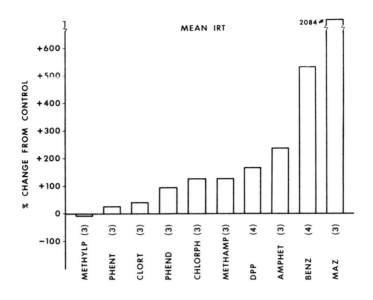

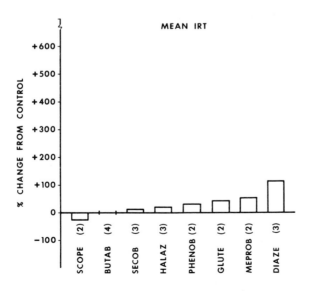

Fig. 2. Average interresponse times obtained on high dose drug days expressed as percent change from control. The number of subjects included in each bar is indicated within the parentheses next to the drug name (complete names in Table 1).

RESPONSE EFFICIENCY

Figure 3 presents an additional dependent variable for the same drugs discussed above. This measure is obtained by calculating the optimum (minimum) number of response required to complete the task at each of the chain lengths and comparing this to the actual number of responses required by the subject. The optimum number of responses required to complete a chain length of six levers is 168. This ratio is obtained for each control day and the average of these days is compared against the average performance following three high doses of any particular drug and expressed, as with the other measures, as a percent change from control.

In considering the results depicted in Figure 3, it may be noted that response efficiency is greater with the stimulant drugs than with the depressant drugs. Thus, even though the average chain length is depressed for subjects receiving stimulant drugs prior to testing, and overall response rates are lower as shown by the increase in mean IRT, those responses made by the subject when it responds tend to be correct. Following depressant drugs, however, neither chain length nor IRT is greatly affected, but efficiency of responding (optimum/actual responses) does decline.

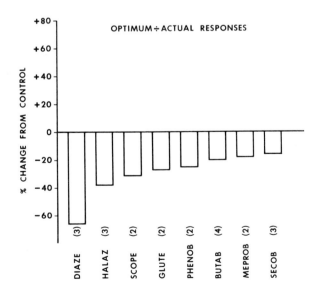

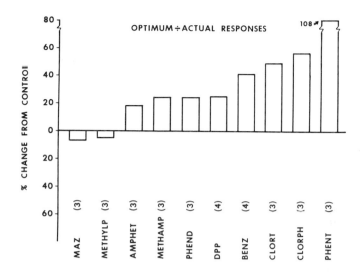

Fig. 3.   Average response efficiency achieved on high dose drug days expressed as percent change from control. The number of subjects included in each bar is indicated within the parentheses next to the drug name (complete names in Table 1).

DISCUSSION

The three dependent measures analyzed thus far suggest that this sequential learning task can yield information regarding the likelihood that a particular drug may produce behavioral toxicity, if administered at greater than therapeutic doses. Although the eight depressant drugs tested thus far produced relatively limited decrements from baseline performance with respect to two of the dependent measures, response efficiency is reduced. In this regard, it seems possible that these drugs are acting to attenuate the consequences of making errors. Specifically, when a subject pulls the incorrect lever, a 7-second timeout period is programmed where the house light is turned off and a buzzer sounds. This contingency can be viewed as punishment which would ordinarily tend to reduce the probability of responding. In the case of depressant compounds, however, the drug state may be interacting with this punishment (McMillan, etc), and, therefore, the animals continue to respond even though they are making errors and receiving more timeout periods (punishment). Because they persist in responding, however, they eventually complete the task. On the

other hand, the interaction of stimulant drugs with punishment
would be more likely to reduce responding than on control days
and the subjects simply quit responding after a relatively few
timeout periods. Thus, they complete the shorter chain lengths
which are presumably easier, but as they progress to the more
difficult longer chains, they begin to make errors and, thus,
stop responding. This could account for the somewhat paradoxi-
cal finding that deficits in the chain length and IRT measure
are accompanied by increased response efficiency with eight of
the ten stimulant drugs tested. More complete answers to these
problems will require further analysis and testing.

In the past, drugs which were not effective orally and
drugs that were judged by the subjects to be noxious tasting
were difficult or impossible to evaluate since, until recently,
the oral route of administration was used exclusively with
these large subjects. Now, however, morphine is being successfully administered by intramuscular injection to two great
ape subjects. Once this procedure has been extended to all six
subjects, the possible range of drugs which can be tested using
this task can be greatly extended.

REFERENCES

Boren, J. J. and Devine, D. The repeated acquisition of behavioral chains. Journal of the Experimental Analysis of
Behavior, 11: 651-660, 1968.
McMillan, D. E. Determinants of drug effects on punished
responding. Federation Proceedings, 34: 1870-1879, 1975.

FOOTNOTE

Portions of the data included in this paper were pre-
viously reported to the Committee on Problems of Drug
Dependence of the Assembly of Life Sciences, NAS-NRC,
April 19-21, 1976.

A SURVEY OF ADVANCES IN CHIMPANZEE REPRODUCTION

C. E. Graham

Yerkes Regional Primate Research Center
Emory University, Atlanta, Georgia

The centennial celebration of Robert Yerkes' birth provides an appropriate opportunity to evaluate advances in
knowledge of chimpanzee reproduction, since research in this
area was an important activity early in the history of the
Yale Laboratories of Primate Biology.

In the book The Great Apes published by Robert and Ada
Yerkes in 1929 (49) the chimpanzee life cycle was reviewed.
The limited available information was largely based upon anecdotal accounts, and lacked accurate data on such basic reproductive parameters as length of the sexual cycle, time of
menstruation with respect to sexual swelling changes, and the
duration of pregnancy.

The development of modern chimpanzee reproduction research may be divided into 3 phases, beginning with the period
1928-1949 during which Robert Yerkes either initiated, or was
closely associated with studies which addressed these questions
and many others. The 20 years from 1950 until 1969 are characterized by a general paucity of publications on chimpanzee
reproduction, particularly by the Yerkes Laboratories. During
the third phase, from 1970 to the present time there has been
a resurgence of interest in the chimpanzee, both at Yerkes
Primate Center and elsewhere. This modern blossoming is
partly due to the availability of new and powerful analytical
tools, and partly due to the availability of more adequate
funding to support the relatively high cost of working with
these animals.

From the inception of his work with chimpanzees, Robert
Yerkes recognized the importance of studies on reproduction
partly because he took the wide view of the value of chimpanzees in biological research. He was also aware of the need to
breed chimpanzees in captivity for medical research, and recognized the need for an adequate basis of physiological know-

ledge. Most important to Yerkes perhaps, was an increasing
awareness of the influence of sexuality on chimpanzee behavior.
        In this review I shall mention only the more important
papers on the subject of chimpanzee reproduction, with emphasis
on physiological studies.
        The first paper on any aspect of reproductive biology to
appear from Yerkes or his associates was by A. C. Bingham in
1928 on the development of sexual behavior in chimpanzees (3).
Soon after the laboratories were established in Orange Park,
Florida, certain aspects of the menstrual cycle were described
by O. L. Tinklepaugh (40-42).
        In 1935, Edgar Allen, well known as a reproductive bio-
logist, and associated with early successes in isolation of
ovarian hormones, published with A. W. Diddle and J. H. Elder,
a study of estrogen (in those days known as theelin) content in
pregnancy urine and placenta (2). The following year these
authors completed a similar study during the menstrual cycle,
and succeeded in demonstrating that the amount of estrogen excreted was roughly correlated with the size of the sexual
swelling in the first half of the cycle (1). A crude chloro-
form extraction technique was used, the first step of which
was putrifaction of urine at room temperature for 10 days. It
is surprising that Allen was able to find as many as 3 other
scientists, Diddle, Burford and Elder, to collaborate with him
on this odiferous study! The final extract was injected into
mice, where the minimal dose which induced vaginal cornifica-
tion was defined as one mouse-unit (this bioassay was known as
the Allen-Doisy test, after its innovators). These first endo-
crinological studies on the chimpanzee may seem crude today,
in the age of chromatography and radioimmunoassay. Yet knowledge of mammalian, and particularly primate reproduction
developed rapidly, to the credit of its pioneers, among whom
Robert Yerkes and his associates should be numbered.
        The first paper on reproduction of chimpanzees by Robert
Yerkes himself, co-published with Dr. J. H. Elder in 1936, was
entitled "The sexual and reproductive cycles of chimpanzees",
appearing in the same year as "Oestrus, receptivity and mating
in chimpanzee" by the same authors (46,47). These two papers
were the first to systematically describe the main features of
the chimpanzee menstrual cycle, including the sexual swelling
changes during the cycle, which have become so valuable as
markers of endocrine function (Fig. 1).

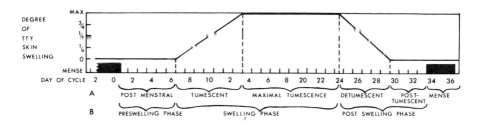

Fig. 1. The phases of the chimpanzee reproductive cycle showing alternative terminologies, A after Yerkes and Elder (47), and B after Young & Yerkes (51). (Figs. 1 and 4 reproduced with kind permission of Karger, Basel).

In 1938, there appeared an important paper by J. H. Elder, "The timing of ovulation in chimpanzee" which by means of timed matings demonstrated that the fertile period was the second half of the period of maximal swelling (11). In 1938, Elder, with Hartman and Heuser of the Carnegie Institute, described a 10 1/2 day chimpanzee embryo which like man, and in contrast to the rhesus monkey, demonstrated interstitial implantation; this embryo was used in the reconstruction of early human development (12).

In 1941, Fish, Young, and Dorfman, the latter known for his lifetime of research on the chemistry of steroid hormones, conducted more advanced extractions of ovarian steroids from menstrual cycle urine, again using the Allen-Doisy assay (15). Although pure compounds were not isolated, fractions were separated containing estrone, estradiol-17β and estriol. They succeeded in demonstrating a follicular and a mid-luteal peak of estrone during the menstrual cycle such as occurs in women; the presence of estriol in the urine, also a characteristic of the human female, was confirmed. Simian primates by contrast do not have a mid-luteal estrogen peak, and do not excrete estriol in measurable quantities. In this study, the excretion of significant quantities of androgen by females was also established, although no temporal correlation with the sexual cycle or with sexual behavior was noted. This subject has not been investigated since in the chimpanzee, in spite of evidence that androgen levels do vary during the cycle in rhesus monkeys, and that they play a role in sexual receptivity. Excretion of androgens and estrogens by male chimpanzees was also demonstrated in this study. Subsequently Fish and Dorfman administered testosterone to a male chimpanzee, demonstrated an increased titer of androgens in the urine, and identified the specific compounds androsterone and etiocholanone (13).

In 1942, Fish, Dorfman and Young isolated pregnanediol from chimpanzee pregnancy urine (14). This compound is the chief metabolite of progesterone in women and was therefore thought to originate in the same way in the chimpanzee: it is not excreted in significant quantities by simians.

In 1937, and 1943, Yerkes and Elder, and Nissen and Yerkes reported on chimpanzee births in the Yerkes Laboratories of Primate Biology (48, 33). The second study summarized 49 births, attesting to the success of breeding the chimpanzee colony. The length of gestation was established at approximately 228 days, and many other aspects of pregnancy, parturition and early post-natal relationships were described.

In 1943 there appeared a landmark paper by Young and Yerkes, in which an exhaustive study of the sexual cycle was presented, with analysis of various factors affecting the duration of the individual phases of the cycle (51). The breadth of this paper has not been surpassed. The study demonstrated longer cycles in the adolescent and postpartum periods and in the winter. The menarche, which occurred at a mean age of 8 yrs. 11 mos. was also affected by a seasonal factor, since it most often occurred in the summer months. The cycle length of the chimpanzee cannot be defined exactly, since many factors influence it, but mean and modal length for 653 cycles were found to be 37.28 and 33 days respectively.

In 1942, Yerkes retired as Director, but for a few years his influence was reflected in a series of studies on sexual behavior continued by his colleagues. In 1944, the first studies appeared on behavioral modification in ovariectomized and ovariectomized-hormone treated chimpanzees, by Young and Orbison (50). A series of related studies followed by Clark and Birch (4,5,7,9). These workers showed that sexual swelling could be abolished by ovariectomy, restored by estrogen administration, and inhibited by progestins. Fluctuations in swelling size occurred with prolonged, constant estrogen treatment, and also in some pregnant chimpanzees. Neither of these phenomena have been explained. Clark found in 1949 that castrate male chimpanzees do not respond to estrogen treatment by developing a swelling, unlike male baboons (6).

At this point we may pause to contemplate the scope of early studies performed by Yerkes and his associates. Shortly after his retirement Yerkes wrote in his book Chimpanzees, a Laboratory Colony (45), "What has been accomplished is trivial by comparison with what waits to be done, but a beginning has been made which should encourage further progress". He was too modest, since the accomplishments of Yerkes and his colleagues in the field of chimpanzee reproduction included documentation of patterns of reproductive cyclicity and pregnancy, correlations of sexual behavior, investigation of hor-

monal control of sexual behavior and sexual swelling, analysis
of the metabolism and excretion of steroid hormones, estimation
of the time of ovulation, and successful breeding, to name the
more important areas.

During the next 20 years no studies of chimpanzee reproductive physiology appeared from the Yerkes Primate Center.
However, in other laboratories a few studies of steroid metabolism were performed which confirmed, amplified and extended
the work of Fish and Dorfman, and Allen and his colleagues.
These new studies were made possible by the advent of purified
steroid hormones, radiolabelled steroids, and improved physio-
chemical separation and identification techniques such as
chromatography and fluorometry. For instances, Romanoff and
his co-workers confirmed the similarity to women in metabolism
of progesterone and 17-OH-progesterone by chimpanzees (37,38).
A similar isotope study by Jirku and Layne confirmed that the
chief urinary metabolites of estrone are estradiol-17β and
estriol (25). The metabolism of testosterone into androsterone
and etiocholanone reported by Fish and Dorfman was confirmed
by administering labelled testosterone to immature chimpanzees (29).

A unique study of reproductive behavior of chimpanzees
in the wild was published by Jane van Lawick-Goodall in 1969
(28). This study contrasted with a paper by Kollar et al., in
1968 describing behavior patterns of a captive group of animals
in New Mexico: this latter study emphasized deficiencies in
sexual patterns of laboratory confined chimpanzees especially
when previously sex-segregated (27). However, more appropriate copulatory patterns could be learned by maturing individuals if they were placed in an adequate social group, thus
supporting the view of Yerkes that coital behavior is mainly
learned by imitation. In 1969, Rogers and Davenport showed
that even if chimpanzees are reared in isolation, their sexual
behavior can to a great extent be rehabilitated by contact with
socially experienced animals (36).

About 1970, a resurgence of interest in chimpanzee reproduction developed at Yerkes Primate Center, and elsewhere. I
first became interested in the subject in 1968, largely because
it dawned on me that the developing availability of modern
displacement analyses for ovarian hormones and possibly gonadotropins, coupled with the increased financial support, parti-
cually for contraception oriented research, offered a great
opportunity for productive research with a species which
earlier studies had demonstrated was of great potential interest.

In 1972 I published in Bourne's comprehensive treatise
on the chimpanzee, an exhaustive study of the microanatomy of
the chimpanzee genital system, demonstrating close similarity

with man (17). In the same year, our group (18) examined the
identify of ovarian hormone urinary metabolites, and their
pattern of excretion during the menstrual cycle more thoroughly
than had been done previously, (Fig. 2), using precise chemical

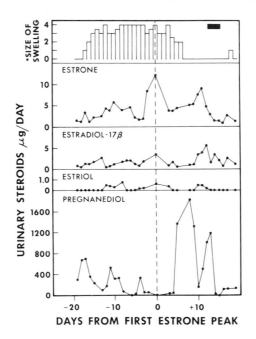

Fig. 2. Sexual swelling and urinary excretion of ovarian
steroids during the chimpanzee menstrual cycle. The day of the
late follicular estrogen peak is designated day 0. (Figs. 2
and 3 reproduced with kind permission of J. B. Lippincott Co.)

assays and rigorous identification procedures; we confirmed the
close similarity to women that earlier workers had reported.
We also reinvestigated the hormonal control of the sexual swelling. Development of sex swelling was associated with estrogen
secretion in the first half of the cycle, or with administration of exogenous estrogens to ovariectomized animals (Fig. 3).
Yet swelling did not develop during the luteal estrogen peak
in the second half of the cycle, suggesting that some other
factor, inhibitory in nature, was controlling sexual swelling.
We found that the luteal phase plasma progesterone and urinary
pregnanediol levels are elevated (18,21). We confirmed that
exogenous progestins could inhibit estrogen-induced swelling
in ovariectomized animals. We therefore concluded that the
absence of swelling in the luteal phase is due to inhibition
by progesterone secreted by the corpus luteum. Thus the

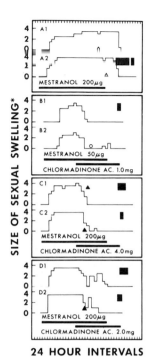

24 HOUR INTERVALS

Fig. 3. Selected examples of sexual swelling pattern during the administration of estrogen (mestranol) and progestin (chlormadenone acetate) in various dose schedules to an ovariectomized chimpanzee. Duplicate experiments (e.g. A1,A2) show replicability of results.

swelling and its detumescence provide a valuable marker for estrogen and progesterone secretion in the chimpanzee and these changes are frequently used by us to time experimental procedures to the appropriate moment of the menstrual cycle.

Although we found that plasma levels of progesterone were similar to the human, urinary pregnanediol levels were considerably lower. The existance of another metabolite of progesterone was therefore suspected, particularly since androsterone is the major metabolite of progesterone in the simian primates. However in a careful study by YoungLai, Graham and Collins (52) using C^{14}-progesterone, this possibility was excluded.

We adapted the technique of endoscopy for use in chimpanzees (Fig. 4 & 5). We are able to make anatomical observations, photographs, ovarian biopsies, inject substances into the ovary as well as aspirate follicles (20). Using this technique we

found only ripening follicles before the last day of maximal
swelling and only corpora lutea afterwards. I conducted an
analysis of endometrial changes throughout the menstrual cycle
which also contributed evidence that the time of ovulation is
closely correlated with the last days of maximal sexual swelling, on the basis that secretory transformation of the endometrium occurs approximately 24 hours after the beginning of
sexual swelling detumescence. Our studies of urinary ovarian
steroids were consistent with this interpretation, and an unpublished study with J. D. Neill of plasma luteinizing hormone
levels provided further strong evidence, since we found that
LH values in 300 chimpanzee plasma samples peaked on the
penultimate and last days of maximal sexual swelling.

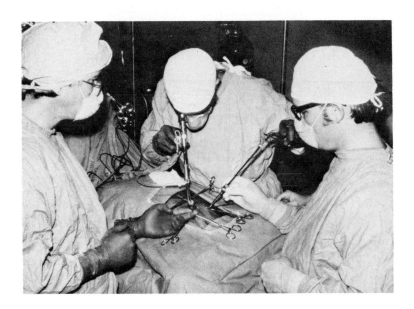

Fig. 4. Examination of chimpanzee ovaries with the
endoscope. Assistants steady accessory instruments.

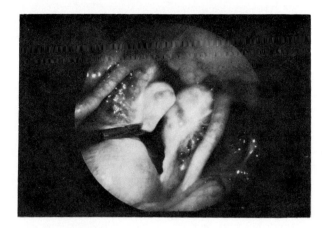

Fig. 5. Reproductive organs of chimpanzee photographed through the endoscope. Ovaries are seen in the center, one of them held by a grasping forceps, on either side are the oviducts, and the fundus of the uterus is at bottom left.

Our study of endometrial development throughout the menstrual cycle revealed close similarities with man (16). Predecidual tissue developed in the uterus of unmated chimpanzees towards the end of the luteal phase; in the rhesus monkey and lower mammals, predecidual tissue develops only if mating occurs and a blastocyst reaches the endometrial cavity. Since man and chimpanzee share the apparently unique features of a luteal estrogen peak and spontaneous predecidual formation, we hypothesized that the mid-luteal estrogen secretion induces predecidualization. We were able to induce a proliferative endometrium in ovariectomized chimpanzees by estrogen administration, and transform the endometrium to the predecidualized state by switching to progestin administration. Therefore, further estrogen administration apparently was not required after the initial period of priming to induce predecidualization. To test this conclusion further, antiestrogens were administered in the luteal phase of intact chimpanzees: predecidualization was not inhibited (19). Thus it may be concluded that the use of antiestrogens to inhibit the decidualization as a means of contraception for women is not a possibility. This analysis could not have been conducted in any species other than the chimpanzee.

Faiman and his colleagues have measured serum ovarian steroids, gonadotropins and prolactin throughout the menstrual cycle and pregnancy (24,35). Their findings corresponded with our menstrual cycle urinary data, except that for some unexplained reason they failed to detect a temporal relation between sexual swelling detumescence and the periovulatory endocrine changes.

These workers have also shown that the endocrine patterns of the chimpanzee during pregnancy much more closely resemble man than do those of simian primates: for example the secretion of chorionic gonadotropin (48) is prolonged in the chimpanzee and woman, whereas it is very restricted in the rhesus monkey (10,34). Progesterone and estrogen levels in the chimpanzee resemble those of women rather than simians (35). These, and direct metabolism studies (39) suggest very similar placental pathways to the human. Immunological pregnancy tests capable of detecting pregnancy as early as 18 days after swelling detumescence have been established in our laboratory as valuable management and research tools for chimpanzees (44).

Recently, close similarities have been demonstrated between chimpanzee and man in placental morphology, and chorionic gonadotropin content (30,22). The immunological and biological properties of chorionic gonadotropin from the two species are also very similar, as shown by parallel dose response of chimpanzee and human chorionic gonadotropin, and by neutralization experiments (23,34). Since a less close antigenic relationship exists between simian and human chorionic gonadotropin, the chimpanzee will probably be the species of choice for testing toxicity of a new generation of contraceptive agents utilizing antiserum to chorionic gonadotropin. Indeed the many similarities in the physiology of pregnancy seem to point to the chimpanzee as the species of choice for studies aimed at human pregnancy.

This extensive work on female chimpanzees is not matched by studies on the male. McCormack has measured testosterone concentration and binding in chimpanzee plasma and found no remarkable differences from the human (32).

The chimpanzee may be a useful subject for the study of male fertility and its control, since the morphology of the genital tract and of sperm is similar to men (17,31), and semen samples can be obtained from chimpanzees by electroejaculation (43) or by suitable training. Using such material obtained at Yerkes, R. V. Short has demonstrated a higher concentration of 19-hydroxyprostaglandin E_1 in chimpanzee semen than in any other ape or man (26): thus the chimpanzee may be uniquely useful for study of the biological activity of this little understood prostaglandin and its possible role in stimulating sperm transport.

This is a year in which we have all been enjoying centennial thoughts of one kind or another. This year is the
tricentennial of the publication by Antonj van Leeuwenhoek of
his famous paper <u>Animacula in Seminae</u>. This paper which appeared in the Philosophical Transactions of the Royal Society
of London, described human sperm for the first time, as seen
through a primitive microscope (Fig. 6A). Martin, Gould and
Warner in our group have recently described the surface morphology of chimpanzee sperm (Fig. 6B) in intimate detail using
the scanning electron microscope (31). Van Leeuwenhoek was
the ultimate male chauvanist; he had no use for the egg, believing that sperm grew directly into an embryo within the
uterus.

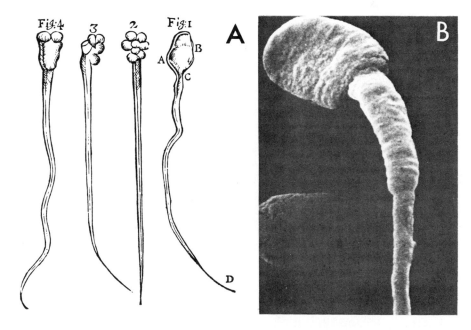

Fig. 6. A, Human spermatozoa as illustrated by
Antonj van Leeuwenhoek. B, Chimpanzee spermatozoa viewed
with the scanning electron microscope.

Nineteen-seventy-five was the bicentennial anniversary of
the demonstration by the Italian physiologist, Lazaro Spallan-
zini that semen is actually required for the development of
vertebrate eggs. This he did by fitting male frogs with tiny
trousers which prevented eggs laid by the female from becoming

"bedewed" with semen, as he quaintly expressed it. He also successfully artifically inseminated a bitch. Two hundred years later, armed with this valuable information, and other more modern data, our team artificially inseminated a chimpanzee on a single occasion, resulting in a pregnancy of 244 days, and the birth of a healthy female 100 years after Robert Yerkes entered the world. I think Dr. Yerkes would have regarded this success as an appropriate celebration of his centennial year, one which argued well for the enhanced productivity of our breeding colony. We hope that by means of artificial insemination, and related measures, Yerkes Primate Center can become totally self sufficient in the production of chimpanzees at a time when chimpanzees may be declared an endangered species, and their importation proscribed.

In conclusion, Robert Yerkes, and those that have followed in his footsteps have explored many facets of chimpanzee reproductive biology, and have discovered many close resemblances to man which are frequently not shared by other laboratory primates. Thus thanks to Robert Yerkes' vision, the chimpanzee has a unique role as an experimental animal in the field of reproduction research.

REFERENCES

1. Allen, E., Diddle, A. W., Burford, T. H. and Elder, J. H. Endocrinology 20: 546-549, 1936.
2. Allen, E., Diddle, A. W., and Elder, J. H. Amer. J. Physiol. 110: 593-596, 1935.
3. Bingham, H. C. Comp. Psychol. Monogr. 5(1): 1-165, 1928.
4. Birch, H. G. and Clark, G. Psychosom. Med. 8: 320-331, 1946.
5. Birch, H. G. and Clark, G. J. Comp. Psychol. Physiol. 43: 181-193, 1950.
6. Clark, G. Yale J. Biol. Med. 21: 245-247, 1949.
7. Clark, G. and Birch, H. G. Psychosom. Med. 7: 321-329, 1945.
8. Clark, G. and Birch, H. G. Bull. Canad. Psychol. Assoc. 6: 13-18, 1946.
9. Clark, G. and Birch, H. G. Endocrinology 43: 218-231, 1948.
10. Clegg, M. T. and Weaver, M. Proc. Soc. exp. Biol. Med. 139: 1170-1174, 1972.
11. Elder, J. H. Yale J. Biol. Med. 10: 347-364, 1938.
12. Elder, J. H., Hartman, C. G. and Heuser, C. H. JAMA 111: 1156-1159, 1938.
13. Fish, W. R. and Dorfman, R. I. Endocrinology 35: 22-26, 1944.

14. Fish, W. R., Dorfman, R. I. and Young, W. C. J. Biol.
    Chem. 143: 715-720, 1942.
15. Fish, W. R., Young, W. C. and Dorfman, R. I, Endocrinology
    28, 585 590, 1941.
16. Graham, C. E. Folia Primatol. 19: 458-468, 1973.
17. Graham, C. E. and Bradley, C. F. In: The Chimpanzee,
    Vol. 5 (G. H. Bourne, ed.): 77-126, Karger, Basel, 1972.
18. Graham, C. E., Collins, D. C., Robinson, H. and Preedy,
    J.R.K. Endocrinology 91: 13-24, 1972.
19. Graham, C. E., Gould, K. G., Wright, K. and Collins, D. C.
    Proc. VIth Int. Congr. Primatol., Cambridge, England,
    Aug. 23-27, 1976, in press.
20. Graham, C. E., Keeling, M., Chapman, C., Cummings, L. E.
    and Haynie, J. Am. J. Phys. Anthropol. 38: 211-216,
    1973.
21. Graham, C. E., Wright, K., Collins, D. C. and Preedy, J.R.K.
    IRCS (Research on: Endocrine System: Physiology; Repro-
    duction, Obstetrics & Gynecology) 2: 1697, 1974.
22. Hobson, B. M. Folia Primatol. 23: 135-139, 1975.
23. Hodgen, G. D., Nixon, W. E., Vaitukaitis, J. L., Tullner,
    W. W. and Ross, G. T. Endocrinology 92 (3): 705-709,
    1973.
24. Howland, B. E., Faiman, C. and Butler, T. M. Biol. Reprod.
    4: 101-105, 1971.
25. Jirku, H. and Layne, D. S. Steroids 5: 37-44, 1965.
26. Kelly, R. W., Taylor, P. L., Hearn, J. P., Short, R. V.,
    Martin, D. E. and Marston, J. H. Nature 260: 544-545,
    1976.
27. Kollar, E. J., Beckwith, W. C. and Edgerton, R. B.
    J. Nerv. Ment. Dis. 147: 444-459, 1968.
28. van Lawick-Goodall, J. J. Reprod. Fert. Suppl. 6: 353-355,
    1969.
29. Layne, D. S., Romanoff, L. P., Forchielli, E., Resnick, O.,
    Kirdani, R. Y., Pincus, G., and Gleason, T. L. (III)
    Aeromedical Research Laboratory Technical Documentary
    Report ARL-TDR-63-25, 1963.
30. Ludwig, K. S. and Baur, R. In: The Chimpanzee, Vol. 4
    (G. H. Bourne, ed.): 349-372, Karger, Basel, 1970.
31. Martin, D. E., Gould, K. G. and Warner, H. J. Human Evol.
    4: 287-292, 1975.
32. McCormack, S. A. Endocrinology 89: 1171-1177, 1971.
33. Nissen, H. W. and Yerkes, R. M. Anat. Rec. 86: 567-578,
    1943.
34. Nixon, W. E., Hodgen, G. D., Niemann, W. H., Ross, G. T.,
    and Tullner, W. W. Endocrinology 90: 1105-1109, 1972.
35. Reyes, F. I., Winter, J.S.D., Faiman, C., and Hobson, W. C.
    Endocrinology 96: 1447-1455, 1975.

36. Rogers, C. M. and Davenport, R. K.   Develop. Psychol. 1: 200-204, 1969.
37. Romanoff, L. P., Grace, M. P., Sugarman, E. M. and Pincus, G.   Gen. Comp. Endocrinol. 3: 649-654, 1963.
38. Romanoff, L. P., Grace, M. P., Sugarman, E. M. and Pincus, G. Gen. Comp. Endocrinol. 3: 655-659, 1963.
39. Shinada, T. and Ryan, K. J.   Steroids 21: 233-244, 1973.
40. Tinklepaugh, O. L.   Anat. Rec. 46: 329-332, 1930.
41. Tinklepaugh, O. L.   J. Morphol. 54: 521-547, 1933.
42. Tinklepaugh, O. L., and van Campenhout, E.   Anat. Rec. 48: 309-322, 1931.
43. Warner, H., Martin, D. E. and Keeling, M. E.   Ann. Biomed. Engin. 2: 419-432, 1976.
44. Woodard, D. K., Graham, C.E., and McClure, H. M.   Lab. Anim. Sci., in press, 1977.
45. Yerkes, R. M.   Chimpanzees. A Laboratory Colony. Yale Univ. Press, New Haven, CT., 321p., 1943.
46. Yerkes, R. M. and Elder, J. H.   Proc. Natl. Acad. Sci. USA 22: 276-283, 1936.
47. Yerkes, R. M. and Elder, J. H.   Comp. Psychol. Monogr. 13 1-39, 1936.
48. Yerkes, R. M. and Elder, J. H.   Yale J. Biol. Med. 10: 41-48, 1937.
49. Yerkes, R. M. and Yerkes, A. W.   The Great Apes; A study of Anthropoid Life. Yale Univ. Press, New Haven, CT, 652p., 1929.
50. Young, W. C. and Orbison, W. D.   J. Comp. Psychol. 37: 107-143, 1944.
51. Young, W. C. and Yerkes, R. M.   Endocrinology 33: 121-154, 1943.
52. YoungLai, E. V., Graham, C. E. and Collins, D. C.   Canad. Fedn. Biol. Soc. Ann. Meeting, 1974.

*SEXUAL BEHAVIOR OF THE CHIMPANZEE IN RELATION TO THE GORILLA AND ORANG-UTAN

Ronald D. Nadler

Yerkes Regional Primate Research Center
Emory University
Atlanta, Georgia

Shortly after the establishment of the Anthropoid Experiment Station at Orange Park, Florida (Yerkes, 1932), Robert M. Yerkes and his colleagues initiated the first experimentally-controlled laboratory studies of sexual behavior and reproduction in chimpanzees.[1] Yerkes' pursuit of this research area in chimpanzees was motivated "primarily by interest in human problems and by the conviction that in its sexual and reproductive life this anthropoid ape is sufficiently manlike, amidst experimentally advantageous differences, to render it peculiarly useful as substitute for man in many types of inquiry" (Yerkes, 1939; p. 78). He believed that, "Through intensive study of the reproductive life of this ape, research leads and methods may be discovered, insights may be achieved, and modes of modification and control developed which will find fruitful application in human social biology" (Yerkes, 1939, p. 78).

In the first study on sexual behavior, conducted in collaboration with James H. Elder, Yerkes sought to define the basic variables that influence mating of chimpanzees, especially regarding the question of estrus or "the regularly recurring appearance or augmentation of sexual receptivity in the female" (Yerkes and Elder, 1936; p. 26). Although limited data on rhesus monkeys (Ball and Hartman, 1935) and chimpanzees (Zuckerman, 1930; Tinklepaugh, 1933) suggested that these

* Preparation of this article and research by the author cited herein were supported by PHS Grant RR-00165 from NIH and NSF Grants GB-30757 and BMS 75-06287.

species did exhibit estrus, other commonly cited reports proposed that the nonhuman primates resembled man, in that they,
unlike lower mammals, mated throughout the sexual cycle
(Hamilton, 1914; Sokolowsky, 1923; Miller, 1928). Yerkes que-
stioned this assertion, not only regarding the nonhuman
primates, but man as well.

Yerkes conducted his investigation of chimpanzee mating
patterns by testing oppositely sexed pairs of individuals
frequently throughout the sexual cycle of the females. He
analyzed the behavioral data in terms of their relationship to
six phases of the female cycle, i.e., menstrual, postmenstrual,
tumescent, maximal swelling, detumescent and premenstrual.

Perhaps the most significant result of this study was the
recognition and description of the complexity in behavioral
interactions of the chimpanzee in even the relatively simple
social setting of the brief paired encounter. On the one hand,
Yerkes asserted with confidence that estrus was indeed characteristic of the chimpanzee cycle. On the other, he cautioned
that no single variable or condition could account adequately
for the variability in behavior or the range of responsiveness
of the various individuals and test pairs he studied.

Regarding the principal question of interest, Yerkes felt
that he had not obtained adequate data to prove the existence
of estrus, since he had not assessed female receptivity in an
independent, quantitative way. However, he inferred the presence of estrus partly from the data on copulation and partly
from analyses of the behavioral interactions and life histories
of the individuals comprising the various test pairs. In terms
of the overall picture, he found that copulation occurred
twice as frequently during the phase of maximal swelling than
during all other phases of the cycle, even though this phase
encompassed only one-third the length of the cycle. Females
initiated mating on 85% of the tests conducted at maximal
swelling compared to only 65% for all the other phases. Conversely, male initiative, less frequent overall, was twice as
frequent during phases of the cycle other than maximal swelling. The data on copulation and female initiative in mating
suggested that female receptivity fluctuated concurrently with
genital swelling and reached a peak in intensity toward the
latter half of the phase of maximal swelling. However, there
were significantly differences between individuals and pairs regarding the period of the cycle during which the males accepted
the females for copulation and regarding the responsiveness of
the females to the males. One male copulated only for a few
days each cycle, during the phase of maximal swelling, whereas
another male copulated frequently and irrespective of cycle
phase. The females, in this respect, differed even more than
the males.

Although Yerkes conducted his investigation in the laboratory, his objective was to describe the typical or characteristic pattern of mating as it would occur in a more natural environment among individuals that had lived together compatibly over long periods of time. He recognized that such conditions of familiarity and compatibility did not exist, for the most part, in his captive subjects. Therefore, he used his personal intimate knowledge of the life histories of the subjects to select as normative examples those pairs which seemed most closely to simulate the relationship of intimacy and congeniality he proposed existed in Nature. By comparing the behavior of these subjects with those whose relationships differed in definable ways, he identified six factor-complexes which he proposed were major determinants of mating. These were described as: 1. physiological, including age, sexual maturation, endocrine status, immediate vigor, etc., 2. Acceptability or attractiveness of one individual for another, presumably determined, in part, by prior experience; 3. the dominance relationship between the individuals; 4. aggressiveness of the male; 5. defensiveness of the female; and 6. extraneous environmental stimuli, primarily visual and auditory.

He proposed, on the basis of such analyses, that under appropriate conditions of familiarity and congeniality, the female chimpanzee determines and controls mating and "The male is suitor and servitor, not lord and dictator" (Yerkes and Elder, 1936; p. 10). On the other hand, when the male is relatively unfamiliar to the female and aggressive and dominant over her, or if the female is immature, inexperienced and/or excessively timid, he "can effect sexual union, if he so desires, whatever the cycle phase of the female and irrespective of her receptivity" (ibid; p. 26). Thus, this initial study, while admittedly deficient in certain respects, provided the first experimental data for elucidating the nature of sexual relationships in a species of great ape. It emphasized the fact that sexual behavior is a social interaction whose occurrence and pattern, in these phylogenetically advanced animals especially, are influenced to a major, if not predominant, degree by the social relations, individual characteristics of the participants and the environmental setting in which the subjects are tested.

As noted above, Yerkes' early conviction that chimpanzees exhibited estrus was not based on objective measures of female receptivity, but primarily on data related to copulation. Since he recognized that "Copulation does not necessarily imply female receptivity" (ibid, p. 38), he designed a second study (Yerkes, 1939) to overcome the deficiencies of the first. The second study was similar in most respects to the first, but in addition, included rating scales to assess quantitatively

female receptivity and the various social factors previously
identified as determinants of mating.

Female receptivity was found to fluctuate cyclically and
concurrently with genital swelling; absent during the non-
swelling phases and at peak frequency during the latter half
of the phase of maximal swelling. The data suggested that the
duration of maximal receptivity was influenced by social relations; relatively brief when the female controlled mating,
but more prolonged when the male was dominant. Acceptability
of the female to the male varied directly with the female's
receptivity and this relationship was consistent for all the
males. Acceptability of the male to the female was also related to cycle phase, but differed for different males. Male
acceptability was directly related to sexual vigor, penis
length, copulation time and the number of copulatory thrusts
and was inversely related to the male's sexual selectiveness,
dominance and aggressiveness. Females approached males most
frequently during maximal swelling when receptivity was high
and copulation most probable. Males approached females most
frequently during the menstrual phase when the females were
nonreceptive and copulation less probable. The males did not
differ significantly in their dominance ratings, but the females differed in timidity and in their acceptance of the
males.   Yerkes found that although copulation generally did
not occur during menstruation, gestation or lactation, it
sometimes occurred during the first 2-3 months of gestation in
relation to genital swelling. He thought that such matings
reflected the male's dominance and his attraction to the
female's genital swelling rather than heightened female receptivity.

In consideration of the significant influence of social
and other psychobiological conditions on chimpanzee sexual
behavior, Yerkes proposed that this species bore a closer re-
semblence to man than any others previously studied. As a
generalization regarding the phylogeny of sexual behavior,
Yerkes (1939) proposed "The higher the order of behavioral
adaptiveness (general intelligence) and the more dominant the
male of the species, the wider the range of copulatory responsiveness in the typical sexual cycle and the greater the tendency of the female to respond accommodatingly to his advances
irrespective of her sexual status" (p. 79). It is appropriate
to note that this hypothesis was proposed when there were few
supporting data for monkeys and humans, and when, in fact, most
of the available evidence on these advanced forms was contrary
to this position. It is also noteworthy that the proposal of
an inverse relationship between species intelligence and hormonal dependence of sexual behavior was expressed also by two
other psychologists who were to become the foremost investi-

gators of animal sexual behavior, William C. Young (Young and
Orbison, 1944) and Frank A. Beach (1942). However, these investigators did not include the qualification regarding male
dominance. As will be shown below, it is this qualification
which makes Yerkes' hypothesis uniquely accurate in characterizing the nature of sexual relations among the great apes,
despite the fact that neither the gorilla nor the orang-utan
had been studied at that time.

A third study of chimpanzee sexual behavior, conducted
at Orange Park by Young and Orbison (1944) had as its objective
determination of the relationship between sexual status,
general body activity and social relations. Of the 18 categories of behavior examined by these investigators, 7 were
found to differ significantly between the follicular phase,
including the preswelling and swelling phases, and the luteal
phase. During the follicular phase, there was a higher frequency of erections prior to formal testing, of females waiting
close to the door to enter the cage of the male, of females
presenting to the males and of copulation, and the subjects
spent a greater amount of time together. An opposite relationship was found for the categories of nonresponsiveness of the
male to the female and of the female to the male. In addition,
data obtained following ovariectomy of three of the females
resembled those from the luteal phase of intact animals and
differed significantly from those of the follicular phase.

Young and Orbison (1944), therefore, confirmed for the
most part, the results of the earlier studies, including the
conspicuous individual and partner differences. They found,
in fact, that individual differences and differences between
partners were statistically significant for all behavioral
categories except presentation by females to different partners. Since only 7 categories of behavior were clearly related
to sexual status, these investigators concluded, as had Yerkes,
that individual and social factors were more important in
determining the outcome of a mating test than was sexual status.
The main difference between the results of this study and the
two previous ones was that Young and Orbison, by analyzing
group averages, failed to find a consistent peak in female
receptivity during the second half of the phase of genital
swelling. They did find peaks in receptivity, however, in the
records of some individuals. This study and the two previous
ones represent the only controlled laboratory studies of sexual
behavior conducted on chimpanzees and were the only such
studies conducted on any of the great apes for more than 30
years. As a whole, these studies suggested that cyclic augmentation of sexual receptivity was characteristic of the
female chimpanzee, that copulation typically occurred for a
relatively brief period during maximal genital swelling, but

that copulation could occur during any phase of the cycle, or
during menstruation, gestation and lactation as a consequence
of varied individual, social and environmental influences.

In recent years, information on sexual behavior of chimpanzees has been obtained for animals living under semi-natural
conditions in the field (van Lawick-Goodall, 1968) and within
social groups in captivity (Kollar, Beckwith and Edgerton,
1968; Tutin and McGrew, 1973). Although these studies supported the essential findings described above for animals tested in pairs, they did report some quantitative differences in
certain aspects of behavior. For example, courtship was more
elaborate in group settings and a somewhat greater percentage
of copulations occurred during maximal genital swelling, e.g.
87% (van Lawick-Goodall, 1968) and 83% (Tutin and McGrew, 1973)
vs. 60% (Yerkes and Elder, 1936) and 73% (Yerkes, 1939).

For present purposes, however, the naturalistic studies
are of interest primarily because of the essentially confirm-
atory data they provided for results obtained in the laboratory, "for the environment of captivity may bring about modifications which only comparable studies of the wild animal
will reveal" (Yerkes, 1932, p. 10). In a more general sense,
the naturalistic studies represent the continuing realization
of Yerkes' pioneering efforts in initiating "field and habitat
studies of the great apes in the tropics (to) supplement the
research of each of the laboratories" (ibid, p. 12). Although
Yerkes, himself, conducted only laboratory studies, he was the
original proponent and sponsor of field research on primates
because he recognized that "a background of reliable knowledge
concerning the wild free individuals is wholly essential for
satisfactory use of captive specimens and for safe interpretation of results which they may yield" (ibid, p. 10).

As noted above, it was more than 30 years following the
laboratory studies by Yerkes and his colleagues before such
controlled studies of sexual behavior were conducted on the
other great apes, gorilla and orang-utan. An early report on
a single female gorilla suggested that this species exhibited
cyclic fluctuations in tumescence of the perineal labia comparable to the more extensive genital swelling of the chimpanzee (Noback, 1939). This report remained the only evidence on
the phenomenon until recently when it was confirmed in a study
with nine female gorillas at the Yerkes Regional Primate Research Center (Nadler, 1975a). The length of the genital cycle
in the latter study was approximately 32 days and compared
closely with lengths of copulatory cycles reported for gorilla
pairs living in zoos (Thomas, 1958; Lang, 1959; Reed and
Gallagher, 1963; Tijskens, 1971; Hess, 1973).

The periodic displays of sexual responsiveness and copulation by the zoo animals in all likelihood represent midcycle
activities mediated by endogenous hormonal events in the

females. It is important, however, to describe the behavioral
and morphological relationships in this species under controlled conditions, for comparative purposes as well as practical
ones. The gorilla, like the chimpanzee (and the orang-utan),
is a member of a taxonomic family closely related to man,
study of which can indirectly advance man's knowledge of himself. The gorilla is also an endangered species whose survival
may be ensured through increased information on its reproductive life and social organization. An initial study was conducted, therefore, similar in design to those of the chimpanzee
in which some 20 oppositely-sexed pairs of gorillas were tested
daily and their behavior assessed in relation to daily measures
of labial tumescence (Nadler, 1975b, 1976). In agreement with
others (Zuckerman, 1930; Noback, 1939), it was hypothesized
that the genital swelling of gorillas reflected to some extent
an action of the female hormone, estrogen. Thus, this morphological parameter provided a useful indirect measure for facilitating interpretation of differentially displayed behavior
patterns.

       The results of the gorilla study confirmed various as-
pects of previous reports based primarily on single pairs of
zoo animals, of which Hess' (1973) was by far the most comprehensive. In addition, it provided the first data on the relationship between sexual behavior and labial tumescence. As
noted in the earlier reports, females were the primary initia-
tors of sexual interactions (Fig. 1). They assertively
approached the males either by backing into them dorso-ventral-
ly or by pulling the males onto them ventro-ventrally as they
lowered themselves onto their backs. Hess (1973) also described a form of "inviting" in which the female extended its arm,
palm downward, toward the male. The females sexually presented
to the males during all degrees of labial tumescence, but they
presented more frequently with successively greater degrees of
tumescence. Copulation occurred rarely during the detumescent
condition and ejaculation not at all, whereas both these
activities were most frequent at maximal tumescence. In contrast to most other mammals, the gorillas used a variety of
positions during copulation, including variations of both the
dorso-ventral and ventro-ventral positions. Dorso-ventral
copulation was by far the most common basic position observed
in this laboratory study and was recently reported to be the
primary position used by gorillas in the wild.(Harcourt and
Stewart, in Press).

       Partner preferences were important in the mating of
gorillas, as had been shown previously for chimpanzees. When
compatible partners were tested, copulation occurred in a brief
1-4 day period of maximal or near maximal labial tumescence.
However, several pairs failed to mate at this frequency, and

Fig. 1. Female gorilla exhibits sexual proceptivity by assertively backing into male.

some failed to mate altogether. In such cases, the absence of mating was attributable to the female's avoidance of the male and the male's apparent lack of assertiveness in soliciting copulation. In the gorilla, therefore, social imcompatibility was represented by infrequent mating, whereas in the chimpanzee, incompatibility often resulted in increased mating unrelated to the phase of the cycle. With incompatible chimpanzees, copulation occurred because the male actively solicited copulation and the female generally responded accommodatingly to the male's advances. The differential effect of partner incompatibility on the frequency and distribution of mating in the cycle of chimpanzees and gorillas thus appears related to species differences in sexual assertiveness of the sexes.

The study of sexual behavior in gorillas revealed several conspicuous differences between this species and the chimpanzees. The gorilla mated during a relatively shorter portion of the cycle than the chimpanzees (12% vs. 33%), the females were more assertive sexually, i.e. more proceptive in current terminology (Beach, 1976), and the males were less aggressive sexually. The gorillas also exhibited considerable variability in their copulatory positions, suggesting significant cognitive

control over behavior commensurate with their advanced degree
of encephalization. On the other hand, the brief and restricted period of mating was similar to the pattern observed in
other mammals with far less forebrain development, e.g. the
rodents. Moreover, the close relationship between mating and
labial tumescence suggested that hormonal influences played an
important role in the regulation of sexual behavior by the
gorillas, despite the complexity of their brains and their relatively superior intelligence. Thus, the study on gorillas
did not support the hypothesis (Beach, 1942; Young and Orbison,
1944) that relatively advanced encephalization and intelligence
per se significantly emancipate sexual behavior from hormonal
regulation.

Information on sexual behavior of orang-utans, as in the
case of gorillas, was, until recently, available only for
animals living in zoos. These brief accounts suggested that
copulation in this species occurred relatively frequently,
without relation to the female's cycle (Fox, 1929; Asano, 1967;
Heinrichs and Dillingham, 1970; Coffey, 1972) and often was
initiated forcefully by the male, despite active resistance by
the female (Fox, 1929; Coffey, 1972). In order to obtain data
on sexual behavior of orang-utans, comparable to those on the
other great apes, four oppositely-sexed pairs from the Yerkes
colony were tested daily during the intermenstrual period of
the female (Nadler, in press). Since female orang-utans do not
exhibit a genital swelling during the menstrual cycle, the behavioral data were analyzed in relation to menstruation.

In confirmation of the earlier reports, it was found that
orang-utans mated on almost every test in which they participated. Moreover, essentially all copulations were initiated
by the males and resisted by the females (Fig. 2). The males
initially pursued the females, wrestled them to the floor and
then positioned the females for ventro-ventral copulation. The
females initially fled, while emitting distress vocalizations,
struggled briefly when caught, but eventually reclined passive-
ly in response to the males' forcefulness. Although most copulations began in the ventro-ventral position, numerous changes
and variations in position were observed during most interactions, including dorso-ventral copulation and copulation
while suspended from the roof of the cage. Single copulations
occurred on almost every test, but copulations beyond the first
occurred most frequently during the middle third of the menstrual cycle. The median duration of copulation was considerably greater in orang-utans than in chimpanzees or gorillas,
i.e. 900 seconds vs. 8 seconds (Yerkes and Elder, 1936; Yerkes,
1939) and 53 seconds (Hess, 1973), respectively. Thus, the
orang-utans, like the gorillas, differed from the chimpanzees
with respect to several behavioral parameters considered above,

Fig. 2. Male orang-utan exhibits sexual dominance by aggressively wrestling female to the floor.

but they differed in a direction opposite to that of the gorillas. The mating of orang-utans occurred during a greater proportion of the cycle than the chimpanzees, essentially co-extensive with the cycle, the orang-utan males were more sexually aggressive than the chimpanzees, and the orang-utan females showed very little, if any, initiative in sexual interactions. That these characteristics of orang-utan sexual relations are not merely an artifact of the laboratory setting is apparent from the findings of recent field studies (MacKinnon, 1974; Rijksen, 1975). Under natural conditions, adult orang-utans are semi-solitary and interact relatively infrequently. When the males do come upon females, however, they generally copulate with them in a manner comparable to that observed in the laboratory.

The data on different patterns of sexual behavior in the great apes provide an opportunity to assess Yerkes' (1939) hypothesis, mentioned above, regarding the regulation of mating in different animal species. Yerkes proposed that across phylogeny, two factors were of primary importance in determining the extent to which copulation occurred during the sexual cycle, i.e., species intelligence and male sexual dominance. He hypothesized that the greater the level of intelligence and the more sexually dominant the male of the species, the lesser the

restriction of copulation in the cycle and, therefore, the
less evidence of behavioral cyclicity. Since the great apes
appear to be fairly comparable in intelligence (Rumbaugh, 1970),
the hypothesis with respect to this group may be simplified such
that only male dominance is considered.

Table 1 shows the relationship between the three species
of apes for male sexual dominance, based on the laboratory
studies described above. Yerkes' hypothesis predicts that the
species in which the male is most dominant (orang-utan) would
exhibit the least restricted pattern of copulation throughout
the cycle, i.e., the least evidence of cyclicity in sexual behavior. The species in which the male is least dominant
(gorilla) would exhibit the greatest restriction of copulation,
i.e., the clearest indication of behavioral cyclicity. Comparison of the rankings of the great apes for male sexual dominance
and cyclicity in sexual behavior (Table 1) illustrates that,
for this group at least, the inverse relationship hypothesized
is obtained. The data suggest that relatively advanced species
intelligence functions in a permissive rather than obligatory
manner with respect to the emancipation of sexual behavior from
hormonal regulation. It follows that, in the great apes, one
or more other variables related to the differential control over
mating by the two sexes were more influential in determining
the frequency of sexual interactions in the cycle.

	Chimpanzee	Gorilla	Orang-utan
Male sexual dominance	2	3	1
Cyclicity in sexual behavior	2	1	3
Female proceptivity	2	1	3
Consort accessibility	2	1	3

Table 1. Socio-sexual relationships of the great apes.
Number 1 indicates the highest rank per category, number 3
the lowest rank, and number 2 intermediate.

It is apparent that Yerkes derived his hypothesis directly from his observations of male sexual dominance in chimpanzees. He reported, for example, that the duration of estrus and copulation in the cycle was relatively brief when the female controlled mating, but more prolonged when the male was in control (Yerkes, 1939). Clearly, male sexual dominance refers to the male's control over mating. Although Yerkes stated his hypothesis in terms of the male's behavior, he could have selected the female's behavior equally well, since, in this context, it bears a reciprocal relationship to the male's. This relationship for the great apes is shown in Table 1, in which the term 'female proceptivity' (Beach, 1976) is used to define the female's control over mating. It can be seen that for the species in which the female exhibits the greatest degree of proceptive behavior (gorilla), the male is least sexually dominant. For the species in which the female shows the least proceptive behavior (orang-utan), the male is the most sexually dominant. The chimpanzee thus occupies an intermediate position between the gorilla and orang-utan with respect to the three variables considered above. As a result of its position vis-á-vis the other apes and the range or variability of its behavior under laboratory conditions, the chimpanzee displayed patterns of sexual interaction common to both the other species. These fortuitous circumstances, combined with Yerkes' critical and insightful analyses of behavior, apparently enabled him to induce the hypothetical principles he proposed.

It is not the purpose of the present paper to attempt a comprehensive interpretation of the behavioral relationships discussed above. However, since, in the apes, behavioral cyclicity is related to the degree to which the male and female control mating, it is appropriate to examine some additional data which provide evidence regarding the evolution of these different patterns of sexual interaction. Under natural conditions, the great apes differ considerably in the extent to which suitable consorts are accessible for reproductive purposes. Reproductive success currently is viewed as the end result of adaptive behavior (Crook and Gartlan, 1966; Mason and Lott, 1976). Therefore, the different strategies by which consorts gain access to each other among the different species of great apes may be considered to have evolved as the result of natural selection. The different patterns of sexual interaction in the apes and, more significantly, the manner in which mating is controlled may be examined in terms of their functional significance for species survival under natural ecological conditions.

Among the apes, gorillas live in the most stable social groups (Schaller, 1963). It may be assumed that the animals are well acquainted with each other and that the males are accessible continuously for impregnating fertile females. Under such conditions, a relatively brief period of proceptive behavior, such as occurs in female gorillas, would be sufficient to communicate to the relevant males that conditions were appropriate for copulation. In the unstable chimpanzee groups (Reynolds and Reynolds, 1965; van Lawick-Goodall, 1968), appropriate males may not be accessible at all times and a more prolonged period of proceptivity/receptivity would provide the female additional time to attract males only infrequently encountered. Communication of its impending fertility by the female chimpanzee presumably would be facilitated further by genital swelling which, among the apes, acquires its greatest dimensions in this species. A relatively prolonged period of attractiveness would also enable a female to express a preference for a particular male with which it could consort during the later, fertile stage of swelling. That chimpanzees in the wild do, in fact, form more or less exclusive consortships and subsequently act in ways to avoid others was reported recently (Tutin, 1975).

Whereas, among chimpanzees and gorillas, females play an important role in the control of mating, among orang-utans, males predominantly control mating. Also, orang-utans in the wild are more spatially dispersed than the other apes and, therefore, consorts are least accessible to each other. The females are less mobile than the males and occupy separate, although overlapping, ranges, a number of which may be contained within the larger range of a male (Rodman, 1973; MacKinnon, 1974; Rijksen, 1975). Communication of its fertile state by a female under such conditions would be ineffective because of the relatively great distances between animals. The absence of a cyclic genital swelling and the limited proceptive behavior in female orang-utans reflect the female's minor role in regulating sexual encounters. Data from the field suggest males frequently attempt to mate with females they come upon and that their initiation of mating is so aggressive as to suggest the interpretation of "rape". This practice of forcefully copulating with females may serve a dominance function rather than a reproductive one (Rijksen, 1975). Since this behavior was observed most commonly in subadult males, it may represent their initial attempts to obtain females of their own (Rodman, 1973). Forceful mating might also reflect a minimal development of social skills due to the male's semi-solitary life-style (MacKinnon, 1974). The finding that multiple copulations occurred more frequently during the midcycle period in the laboratory study suggests also that the stimuli associated with initial copulations may communicate to the male information regarding the sexual status of the female.

Although the specific function of the male orang-utan's sexual aggressiveness remains controversial, that males rather than females take the initiative in sexual interactions is clear. The assumption of sexual initiative by the male appears to be a result of its greater mobility than the female under conditions of low consort accessibility which require periodic contact for successful reproduction (Rodman, 1973). Table 1 shows the relationship between consort accessibility for the great apes in relation to the other variables discussed. It can be seen that for the species in which consorts are most accessible (gorilla), the female controls mating and the duration of sexual activity in the cycle is relatively short. For the species in which consorts are least accessible and the male relatively more mobile than the female (orang-utan), the male controls mating and sexual activity occurs throughout the cycle. It is suggested that among the great apes, the degree to which copulation is restricted during the sexual cycle is directly related to the degree that the female controls sexual interactions and inversely related to the degree that the male controls such interactions. Furthermore, the degree to which sexual interactions are differentially controlled by the two sexes is related to consort accessibility in the natural habitate, i.e., the more accessible the consorts, the more the female controls mating. Finally, accessibility of consorts is related to group social structure which is viewed as having evolved in response to specific ecological conditions.

It is especially appropriate to include the comparative assessment of great ape sexual behavior in the Robert M. Yerkes Centennial Conference because this research issue is one to which Yerkes made original contributions and within which his influence is still apparent. Yerkes was one of the first investigators to study the great apes under controlled laboratory conditions and he conducted the first comprehensive investigation of sexual behavior in one of the species of great apes. On the basis of his work with the chimpanzee, Yerkes proposed a broad hypothesis regarding the regulation of mating in the sexual cycle which has proved recently to apply rather precisely to the sexual relations of the gorilla and orang-utan. Yerkes also was one of the earliest proponents of field research as a method complementary to laboratory research for facilitating interpretation of behavioral data. The current resurgence of field and laboratory research on the great apes may be viewed as an affirmation of Yerkes' confidence in the comparative approach for the study of human problems and the scientific use of the apes for "the extension of knowledge of life and the improvement of its quality" (Yerkes, 1939, p.78).

REFERENCES

Asano, M. (1967). Int. Zoo Yearb. 1, 95-96.
Ball, J. and Hartman, C. G. (1935). Amer. J. Obstet. Gynecol. 29, 117-119.
Beach, F. A. (1942). Psychol. Bull. 39, 200-226.
Beach, F. A. (1976). Horm. Behav. 7, 105-138.
Coffey, P. F. (1972). Ann. Rep. Jersey Wildlife Preservation Trust, 15-17.
Crook, J. H. and Gartlan, J. S. (1966). Nature 210, 1200-1203.
Fox, H. (1929). J. Mammal. 10, 37-51.
Hamilton, G. V. (1914). J. Anim. Behav. 4, 295-318.
Harcourt, A. H. and Stewart, K. J. (in press). In "Recent Advances in Primatology" (D. J. Chivers, ed.), Vol. 1 Academic Press, London.
Heinrichs, W. L. and Dillingham, L. A. (1970). Folia Primat. 13, 150-154.
Hess, J. P. (1973). In "Comparative Ecology and Behaviour of Primates" (R. P. Michael and J. H. Crook, eds.), pp. 508-581. Academic Press, New York.
Kollar, E. J., Beckwith, W. C. and Edgerton, R. B. (1968). J. Nerv. Ment. Dis. 147, 444-459.
Lang, E. M. (1959). Int. Zoo Yearb. 1, 3-7.
MacKinnon, J. (1974). Anim. Behav. 22, 3-74.
Mason, W. A. and Lott, D. F. (1976). Ann. Rev. Psychol. 27, 129-154.
Miller, G. S., Jr. (1928). J. Mammal. 9, 273-293.
Nadler, R. D. (1975a). Anat. Rec. 181, 791-798.
Nadler, R. D. (1975b). Science 189, 813-814.
Nadler, R. D. (1976). Arch. Sex. Behav. 5, 487-502.
Nadler, R. D. (in press). In "Recent Advances in Primatology" (D. J. Chivers, ed.), Vol. 1. Academic Press, London.
Noback, C. R. (1939). Anat. Rec. 73, 209-225.
Reed, T. H. and Gallagher, B. F. (1963). Zool. Gart. (Leipzig) 27, 279-292.
Reynolds, V. and Reynolds, F. (1965). In "Primate Behavior, Field Studies of Monkeys and Apes" (I. DeVore, ed.), pp. 368-424. Holt, Rinehart and Winston, New York.
Rijksen, H. D. (1975). In "Contemporary Primatology" (S. Kondo, M. Kawai and A. Ehara, eds.), pp. 373-379. Karger, Basel.
Rodman, P. S. (1973). In "Comparative Ecology and Behaviour of Primates" (R. P. Michael and J. H. Crook, eds.), pp. 171-209. Academic Press, New York.
Rumbaugh, D. M. (1970). In "Primate Behavior. Developments in Field and Laboratory Research". (L. A. Rosenblum, ed.), Vol. 1, pp. 1-70. Academic Press, New York.

Schaller, G. B. (1963). "The Mountain Gorilla". Univ. of Chicago Press, Chicago.

Sokolowsky, A. (1923). Urol. Cutan. Rev. 27, 612-615.

Tinklepaugh, O. L. (1933). J. Morphol. 54, 521-547.

Thomas, W. D. (1958). Zoologica 43, 95-104.

Tijskens, J. (1971). Int. Zoo Yearb. 11, 181-183.

Tutin, C. E. G. (1975). In "Contemporary Primatology" (S. Kondo, M. Kawai and A. Ehara, eds.), pp. 445-449. Karger, Basel.

Tutin, C. E. G. and McGrew, W. C. (1973). Amer. J. Phys. Anthropol. 38, 195-200.

van Lawick-Goodall, J. (1968). Anim. Behav. Mongr. 1, 161-311.

Yerkes, R. M. (1932). Comp. Psychol. Monogr. 8, 1-33.

Yerkes, R. M. (1939). Hum. Biol. 11, 78-111.

Yerkes, R. M. and Elder, J. H. (1936). Comp. Psychol. Monogr. 13, 1-39.

Young, W. C. and Orbison, W. D. (1944). J. Comp. Psychol. 37, 107-143.

Zuckerman, S. (1930). Proc. Zool. Soc. Lond., 691-754.

FOOTNOTE

[1] Throughout this paper all references to chimpanzees apply to Pan troglodytes. Although the pygmy chimpanzee (P. paniscus SCHWARZ, 1929) is of considerable interest, it has not been studied sufficiently in the laboratory or in the field for comparison in this article with the other great apes.

*VISUAL CHARACTERISTICS OF APES AND PERSONS

Francis A. Young and George A. Leary

Primate Research Center
Washington State University
Pullman, Washington

Since the purpose of this session is to consider the chimpanzee as a bio-medical model for the human, most of the discussion involved in the present paper will deal with the characteristics of chimpanzees and persons. While the classification of apes generally includes gorillas, chimpanzees, orangutans and gibbons, in our series of studies we have never been in a position to obtain visual measures on gibbons, although we do have measures on orangutans and relatively young gorillas, which will be included in this paper.

The basic visual similarity of the optical systems of the sub-human primate, including the apes and the human subject, have been detailed elsewhere (Young, 1973) and will be reviewed only briefly here. In any evaluation of animal subjects as substitutes for persons, the greater the defined similarity between the animal subject and the person, the more confidence one has in making generalizations from the characteristics of the animal to the characteristics of the person. Conversely, the more deviations one finds between the characteristics of the two groups, the less confidence one has.

We hope to detail in the present paper, some very recent similarities which we have found in growth changes of a cross-sectional nature in chimpanzees as they relate to similar growth changes in persons. At the same time we will provide additional information about these basic similarities of the orangutan and gorillas with·respect to both the chimpanzee and the human.

*This investigation was supported in part by U.S. Public Health Service research grant EY 00284 from the National Eye Institute; the 6571st Aeromedical Res. Laboratory, Holloman Air Force Base, New Mexico and by the Yerkes Regional Primate Research Center, Atlanta, Georgia.

BASIC STRUCTURE OF THE EYE

1. Characteristic of the globe

All human and subhuman eyes are characterized by three
layers of the membrane or tunic: the outermost white sclera;
a middle vascular-muscular layer, the choroid; and the inner-
most neurosensory layer, the retina. These layers surround
and are distended by the large vitreous gel which fills the
anterior chamber. The normal intraocular pressure varies
between approximately 10 mm of mercury in certain monkeys, to
17 to 18 mm of mercury in persons. A smooth-muscle system in
the ciliary body and the choroid, control the shape of the
lens of the eye, which serves to focus the eye for objects at
different fixation distances. The lens itself is suspended by
radial fibers (the zonular fibers) between the anterior chamber
and the vitreous chamber, approximately 3 1/2 mm from the front
surface of the cornea to the front surface of the lens.
The eyes of human and subhuman primates are also characterized by a clear cornea, which has a considerably smaller
radium than the globe of the eye itself, and is located in the
front of the globe, so that light entering the eye enters
through the cornea. The cornea serves as the major refractive
element in the eye, and differs in radius and power among the
different primates, with humans having the largest radius and
the lowest power, while among the apes it is likely that the
gibbon has the shortest radius and highest power of the cornea.
The actual anatomical structure of the cornea is the same for
all primates.
The iris, located immediately in front of the lens, controls the amount of light entering the eye through an opening
in the center of the iris, the pupil. Under very high levels
the "pin hole" pupil improves the optical quality by restricting the entering light to the very center of the lens area
and by giving rise to a point source of light.

2. Gross retinal characteristics

The apes and persons are equipped with a spot fovea, an
area of approximately 1 1/2 degrees in diameter, containing
tightly packed cone receptor cells, which mediate the maximum
visual acuity or resolving power of the eyes. The foveal area
provides the equivalent of a 1:1 projection of the retinal
image to the visual cortex of the brain. The fovea contains
only the color-receptor cones, which can respond to all colors
as well as to black and white, but do not function in low
levels of illumination, while the rod receptor cells still
function, since they are more sensitive to light than are the
cones.

An ophthalmoscopic examination of the retina, referred
to as a "fundus examination" with an ophthalmoscope or fundus
camera, of various primate eyes, reveals the basic similarity
of the distribution of the arteries within the eye, of the
optic disc and foveal area, and other shape, size and characteristics of the retinal area of the eye.

3. The microretinal characteristics

The microstructure of the retina of human and chimpanzee
eyes has been studied by Polyak (1957), who demonstrated that
the eyes of these subjects are extremely similar in distribution of rods and cones: a rod-free fovea and a layered structure of sensitive receptor cells (rods and cones) in the outmost
layer of the retina, lying close to the choroid itself. Thus
light must pass through all the other layers of the retina
before reaching the sensitive elements which face toward the
back of the eye rather than toward the lens of the eye as might
reasonably be expected. The foveal type of retinal system
found in the primate requires that the eye be capable of a wide
degree of movement in order to bring visual stimuli onto the
foveal area of the retina.

BINOCULAR CHARACTERISTICS

1. Visual Fields

The forward-placed eyes of the human and the apes provides for virtually complete overlap of the visual fields in
the two eyes. Actually, the eye placement and the shape of
the nose and head in the apes allows for an overlap which is
more comparable to the Mongolian eye position than to the
Caucasian eye position, since in the latter the high nose
bridge restricts the nasal visual field. Cowey (1963), has
developed a method of perimetry for use with monkeys and other
subhuman primates, and Cowey and Weiskrantz (1963), carried
out a biometric study of visual-field defects in monkeys. They
found that the monkey responds in the perimetry situation
essentially the same as the human, although they were not able
to obtain measurements as far out in the periphery of the eye
in monkeys.

2. Ocular movements

The subhuman primates show the characteristic humanoid
ocular movements of convergence upon viewing an approaching
object, divergence on viewing a receding object, conjunctive
movements in any direction, as well as rotation movements in-

Further, the relationship between the area of the fovea in area 17, the striate area, is greatly reversed so that the cortical area may be as much as 10,000 times larger than the fovea area in the retina itself (Talbot and Marshall, 1941). Thus, this extremely small part of the retina provides a tremendous amount of the visual information received by the primates.

The nature of the visual response, to a great degree, is determined by the light pattern received on the retina, and strongly illuminated parts of the retina dominate over weakly illuminated parts of the retina, and may give rise to what have been called on and off responses. Further, if the retina is effectively immobilized with respect to the light source, the retina tends to adapt to the light source and ceases to respond (Riggs, et.al, 1953). Or, if a single retina is exposed to a completely uniform visual field, again the retina will adapt to this visual field and ceases to respond, (Albee and Young, 1976). The eyes themselves have micromovements of less than 1/2 minute of arc with frequencies that seem to average around 50 per second and serve to continually vary the stimulation of the sensitive elements in the retinas of the eyes to prevent the fading of the visual image.

THE NATURE OF THE OPTICAL IMAGE

Since the intact retinal-neural system can transmit whatever pattern is received on the retina by way of the optic nerves, the optic chiasm, the optic tract, the lateral geniculate body, and the striate area of the visual cortex, we will consider in more detail the optical characteristics of the eye, of the role that it plays in developing the visual image on the retina, and the modification of this visual image as it occurs in the various apes and in persons.

SUBJECTS

The subjects involved in this study varied in terms of age, sex, numbers, subspecies, and other characteristics, but almost entirely represent a cross-sectional approach as far as the present data are concerned. While we are carrying out a number of longitudinal studies on chimpanzees and humans, the present results are based primarily on cross-sectional measurements in order to achieve the various age levels desired, and comparability with the gorilla and orangutan populations which were seen only once. The ages of the humans from birth records, and of most of the chimpanzees, were accurately determined, either by virtue of the original size and longevity

within the laboratory, or birth records in the case of most of
the chimpanzees, which came from the colony at Holloman Air
Force Base, New Mexico. Unfortunately, while both the gorillas
and the orangutans were measured here at the Yerkes Regional
Primate Research Center we were not able to get accurate age
information on either of these groups, so we have resorted to
the weight curves for larger apes and man, which are presented
as Figure 10 in Reynold's book on the apes (1967). Thus, the
age figures on the gorillas and orangutans are estimates,
whereas those on the chimpanzees and humans are accurate.

Table I presents the numbers of each of the subject
groups by age levels and by sex. In all cases, when a particular refractive condition is selected, such as emmetropia,
(no refractive error), males generally have flatter corneas
with a greater corneal radius, deeper anterior chambers, approximately the same thickness of the lens, lower lens equivalent powers, larger depths of the vitreous chamber, and larger
axial lengths, as well as lower total powers of the eye. These
relationships apply across all apes and humans, but only so
long as the refractive errors are held relatively constant.
Even in the situation where refractive errors are held constant,
the components of refraction differ widely from eye to eye, so
that in an individual male or female eye, there can be a reversal of the component relationships. With the same refractive error, the male may be exceeded in the case of some
components by the female, and exceed the female in the case of
other components. This is particularly true when the corneal
power in the case of the male is high with respect to the
corneal power in the case of the female. In this situation,
the male will have a narrower vitreous chamber depth and a
shorter axial length than the female. Of course, if refractive
error is ignored, then any relationship may be maintained between the male and female eyes.

However, the basic relationships found among humans with
respect to male-female relationships hold equally well for the
apes. These relationships may be briefly summarized as follows:
1. The female eye being smaller, generally tends to
have a higher power in all of its components except vitreous
chamber depth and axial length than the male eye.
2. The female tends to develop myopia earlier than the
male, and thus begins to show an enlargement of the vitreous
chamber and a corresponding increase in axial length.
3. The earlier myopia begins, the greater the amount of
myopia developed, and as a consequence, female subjects in all
groups tend to have higher amount of myopia than male subjects.
4. The development of myopia is closely tied to the
increase in depth of the vitreous chamber, which in turn determines the increase in axial length.

Any development of myopia predisposes toward detachment
of the retina, and we have observed detached retinas in chimpanzees as well as in humans.

In dealing with comparative data between persons and
apes, one must deal with the relative developmental and chronological ages of the species involved. Since most of the three
apes considered, chimpanzees, gorillas and orangutans, show
male-female sexual development at approximately seven to eight
years of age, while the human subjects show male-female sexual
development at between twelve and fourteen years of age, the
ape reaches the same developmental stage in about 60% of the
time required for the human. A comparison of the development
of the optical characteristics of the chimpanzee and the human,
using chronological age, appeared to be more meaningful and
more similar between groups than by using developmental age.
In the figures presented, chronological ages are employed rather
than developmental ages. In fact if one attempts to use a
developmental factor of approximately 1.75 to bring the animal
developmental age in line with the human chronological age,
very peculiar results are obtained.

There are two groups of human subjects used in these comparisons, one is based upon a population of Warm Springs Oregon
Native American children, and the other is based upon a British
population of Caucasian children, living in the poorer areas of
London. These two human populations were used to determine the
sex differences between males and females. We attempted to use
the right eye only, and succeeded in 96% of the cases; occasionally, for some measures, no right eye values were available,
and in this case the left eye value was used in order to maintain the number of subjects.

PROCEDURES

As will be noted in Table I, we were able to obtain
satisfactory measures on chimpanzees and gorillas between 2 and
2.9 years of age, but the youngest human subjects we were able
to evaluate were between 4 and 4.9 years of age. It would have
been possible to obtain comparable measures to those obtained
on the chimpanzees and gorillas if we had been able to use the
same techniques on the human subjects.

TABLE 1.

Numbers of right eyes (or left eye if right eye data is missing) by groups, sex and age levels.

	Ages-Years															Totals
	2.5	3.5	4.5	5.5	6.5	7.5	8.5	9.5	10.5	11.5	12.5	13.5	14.5	15.5	16.5	
CHIMPANZEES																
Females	4	15	26	22	25	19	20	14	6	4	6					161
Males	3	19	23	29	32	29	22	14	11	3	2					187
Combined	7	34	49	51	57	48	42	28	17	7	8					348
GORILLAS																
Females	4	2		1	2											9
Males	2	1		1	0											4
Combined	6	3		2	2											13
ORANGUTANS																
Females					2	3		2	2			1			1	11
Males					0	1		2	1			3			3	10
Combined					2	4		4	3			4			4	21
HUMANS																
British																
Females				19	20	26	44	35	32	30	17	17	21			261
Males				22	29	39	34	30	26	26	22	11	9			248
Combined				41	49	65	78	65	58	56	39	28	30			509
Native Americans																
Females			6	11	10	14	15	20	19	13	9	5	4			126
Males			6	12	20	16	27	22	21	10	10	5	4			153
Combined			12	23	30	30	42	42	40	23	19	10	8			279

On all of the ape subjects, the subjects were anesthetized with Serynlan (phencyclidine hydrochloride) which in approximate 10% of the animals was supplemented with Nembutal. Since anesthetics are used in young children only under special conditions, we were not able to use these techniques on the children, but instead had to rely upon cooperation. The children were more cooperative beginning with age 5 and above, than they were below age 5, therefore, we have relatively few results on young children.

Incidentally, Sernylan has different effects on different groups of primates, and also different effects on different ages of primates. Generally speaking, since Sernylan works as a depressant upon the central nervous system itself, many of the lower order reflexes remain in operation even though the animal is anesthetized, particularly, the eyes remain open, and are somewhat immobilized by the drug. As a matter of fact, it was originally thought that the drug would be excellent for eye surgery in humans, since it produces an almost complete immobilization of the eyes for a considerable period of time, but it also induces psychoactive changes which are disturbing to many people, and consequently it has never been approved for use in humans.

This differential reaction on primates and other animals appears to be related to the relative importance of the cerebral cortices in these animals, since it has the greatest effect in humans, next greatest in gorillas, then chimpanzees, pigtail macaques, rhesus macaques, orangutans, stumptailed macaques, cynomolgous, and so on down the lower primate scale. It causes a great deal of excitement, but little or no anesthetia in dogs, cats, or pigs, but is a very potent anesthetic in elephants and dolphins. It has virtually no effect in the birds. This differential position of the orangutan is worthy of further investigation.

In all of our investigations on humans and apes, we attempted to determine the curvature of the front and rear lens surfaces, by means of phakometry. This technique requires that two lights be directed into the eyes so that the images of these lights are reflected from the front surface of the cornea (the first Purkinje image), from the rear surface of the cornea (the second Purkinje image), from the front surface of the lens (the third Purkinje image), and from the rear surface of the lens (the fourth Purkinje image). These reflected images are photographed and compared to the first images, which are based on the known and measured curvatures of the cornea as determined by keratometry. In order to obtain these Purkinje images, it is necessary to have a widely dilated pupil, and for our longitudinal studies, no accomodation of the lens. To obtain this in all of our subjects, we routinely used various concentrations and dosages of Cyclogyl (cyclopentolate hydro-

chloride) in order to obtain a state of complete cytoplegia.
The accuracy of our techniques can be evaluated by the effectiveness of our phakometric techniques. The use of Ocniylan
on the orangutans was so erratic and out of line with what we
usually got on Chimpanzees and on the gorillas, as well as on
pigtails and rhesus, that we were completely unable to ever
get a phakometric measurement on the orangutan, in spite of the
use of equipment which readily permitted us to get these measurements on chimpanzees. For this reason, we believe that the
orangutan is not comparable in terms of its relative cerebral
dominance with either the chimpanzee or the gorilla, and for
that matter with some of the lower primates.

        In addition to keratometry for determination of the
corneal curvature, and phakometry for the determination of the
lens surface curvatures and powers, we routinely carry out
ultrasonography for the determination of the distances within
the eyes, a cycloplegic refraction to determine the refractive
errors of the eyes, and tonometry to determine the intraocular
pressure of the eyes of all of our subjects. When all of the
measures, including the phakometric measures are available for
the individual eyes, an optical equation of the eye can be
developed, which provides a great deal more information than
can be provided with the usual clinical technique of simply
determining the refractive error through retinoscopy, and the
use of a subjective refraction in the human subject. The
simple determination of refractive error is not comparable to
the complex optical equation of an individual eye.

RESULTS

        The results presented in Figures 1, 2, 3 and 5 are for
all four groups, chimpanzees, gorillas, humans and orangutans,
with males and females combined. In Figure 4, the results are
presented for the males and females separately, as well as combined.
        The vertical ocular refraction (VOR) represents a re-
solution of the measured axes of the cornea into the vertical
direction so as to make the results comparable to the results
obtained by phakometry, which is also operated in only the
vertical direction. This resolution is somewhat similar to the
equivalent spheres approach, which makes use of one half of the
astigmatic correction, and adds it to the spherical correction
for a single statistical value. In the VOR approach, the axes
are actually taken into account, as well as the magnitude of
the astigmatic correction.
        Figure 1 shows the changes occurring in the VOR with
age, over the age range for different groups from 2 1/2 years

to 16 1/2 years.

Figure 2 presents comparable data based on the depth of
the anterior chamber and lens thickness for these same groups,
over the same age range, in which males and females have been
combined. As indicated previously, lens thickness tends to be
very close for males and females, but the depth of the anterior chamber is generally larger for males than for females.
Consequently, one could expect slightly more variability in
the combined results for the depth of the anterior chamber than
one would expect for lens thickness. Since one age group may
be dominated by males, while the next age group is dominated by
females, there can be a considerable variation, which is simply
due to the sex difference rather than to any other difference.
Again, the chimpanzee and human populations are large enough
to provide relatively stable results.

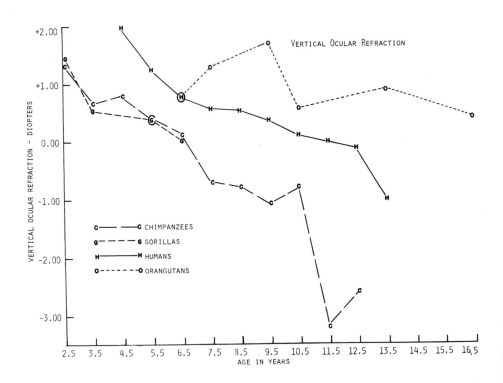

FIG. 1 V-O Refractions for Chimpanzees, Gorillas, Orangutans
and Native American Children

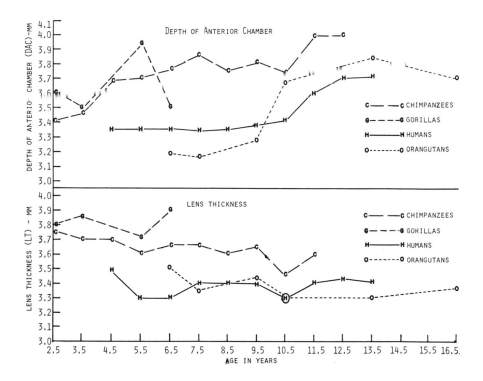

FIG. 2   DAC and LT for Chimpanzees, Gorillas, Orangutans,
and Native American Children

Figure 3 is a plot of the lens equivalent powers for all
of the groups across the age levels, with the males and females
combined.   Again, there are sex differences which are ignored
simply because of the size of the populations involved.   The
lens equivalent power is the power of a single thin lens, in
a predetermined location, in the general area of the crystaline
lens of the eye, which would provide approximately the same
power as does the crystaline lens.   As such, it gives an accurate idea of changes in the crystaline lens power.

Figure 4 presents the same lens equivalent power data
for the Warm Springs Native American children, with the males
and females separately and as a combined group.

Finally, Figure 5 presents the combined males and females for all four groups, in terms of the depth of the vitre-
our chamber and the axial lengths, as measured by ultrasonography.

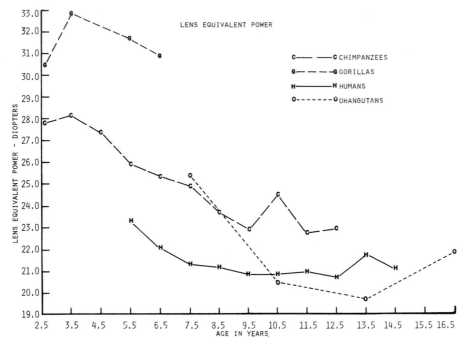

FIG. 3.   Lens Equivalent for Chimpanzees, Gorillas, Orangutans
and British Children

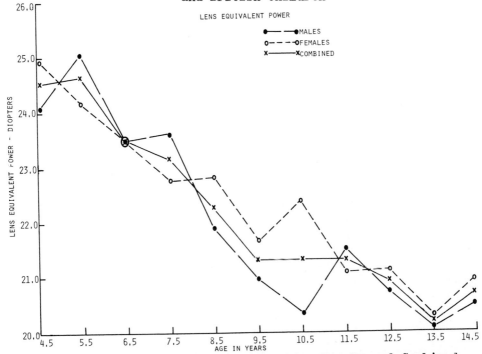

FIG. 4.   Lens Equivalent Power for Male, Female and Combined
Native American Children

Finally, Figure 5 presents the combined males and females for all four groups, in terms of the depth of the vitreous chamber and the axial lengths, as measured by ultrasono graphy.

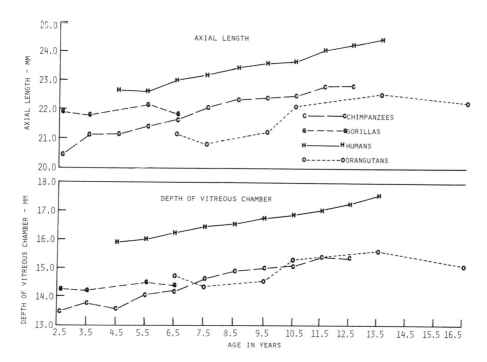

FIG. 5. Axial Lenghts and Vitreous Chamber Depths for Chimpanzees, Gorillas, Orangutans and Native American Children

DISCUSSION

The vertical ocular refraction is quite similar for both males and females, except for the earlier development of myopia, and the greater amount of myopia developed in females. However, the shapes of the distributions are similar, so that no serious effect results in combining males and females as long as the number of males and females is approximately the same. As indicated in Table I, this essentially occurs for both the chimpanzee and the human groups, but does not readily hold for the gorilla or orangutan groups. A very close approximation between the VOR's of the gorillas and chimpanzees is notable, but because of the sample size of the gorillas,

this may be a chance similarity. The basic similarity of the
chimpanzee and human groups is probably not based on chance
because of the number of subjects involved in these two groups.

     It is possible to develop regression equations for the
human and chimpanzee groups, as well as for a variety of other
human populations, which clearly indicate the changes which
occur in refractive conditions over time. All such regression
equations are negative in that the subjects move from higher
plus lenses, or hypermetropia, towards zero, or emmetropia, and
into myopia or use of minus lenses. In the case of the chimpanzee group, the very rapid drop occurring at age 10 1/2 was
primarily due to female subjects. Whether or not an individual
subject actually becomes myopic will be determined by the slope
of the regression line, and the amount of hypermetropic reserve
with which he begins. The equation for the human line is
Figure 1 is $Y = +2.00 -.30X$, where X is the yearly age incre-
ment, beginning with age 4.5. Thus a person with a +2 diopter
refractive error, progressing at the average rate of -0.30
diopters per year over a ten year period, would end up with 1
diopter of myopia at the end of that time, but if he began with
3 diopters plus, over the same time span, he would be emmet-
ropic if he progressed at the average rate.

     The regression slopes are quite different for different
samples, and depend upon the characteristics of the subjects'
behavior more than upon any other variable. Since those subjects with a great exposure to a near-point environment show
slopes as high as -0.50 or -0.60 diopters per year, while those
with very low exposure to near-point environment may show
slopes as low as -0.10 or less diopters per year.

     Figure 2 shows the changes occurring in the anterior
chamber between 2 1/2 and 16 1/2 years of age for the different groups. As indicated, the orangutan and gorilla groups
are so small in number that the values of different years reflect the difference in sexes more than they represent actual
changes occurring in the depth of the anterior chamber.
However, both the chimpanzee and human curves are based on
samples large enough to be relatively stable and reliable indicators of the changes which are occurring. As can be seen
in both groups, for comparable age levels between 4 1/2 and
10 1/2 the slopes of the lines are relatively the same. In
both groups, however, at 10 1/2 there appears to be an increase
in the depth of the anterior chamber. This increase in anterior chamber depth is a characteristic developing myopes, which
has been noted many times by clinicians. The development of
myopia begins with a change in the equivalent power of the lens
and an increase in the depth of the anterior chamber, which is
then followed by an increase in the depth of the vitreous
chamber. It is clear that the variability from year to year

in the depth of the anterior chamber is considerably greater than the variability from year to year in the case of lens thickness.

Again, however, the lens thickness for humans and chimpanzees, as well as for gorillas and orangutans, are quite comparable, showing very little variation in terms of actual magnitude, since the entire range is only one millimeter from year to year in any of the groups. Generally speaking, within the age ranges shown, the lens thickness remains relatively constant. However, the lens in the eyes of both apes and humans continues to grow throughout the life span of the individual, and as a result of this, becomes thicker after the individual reaches the adult age level. Thus, it is not uncommon for an individual who has a lens thickness of 3.5 millimeters at age 10, to have a lens thickness of 4.5 millimeters at age 50. In general, lens thickness does not contribute a great deal to the changes in refractive characteristics or in the development of myopia.

Figure 3 represents the changes in lens equivalent over the years, and it will be noted that the human population in this Figure is different than the human populations in Figures 1, 2, 4 and 5. In Figure 3 the human group comprised a sample of British children who lived in an economically depressed, or ghetto, area of London. These children were poorly educated and not academically motivated, so that the changes which occurred in lens equivalent power were similar in both males and females, and the combined curve accurately represents the changes for both groups. It is also noticeable that the total change varied from approximately 23 1/2 diopters at age 5 1/2 to slightly less than 21 diopters at age 12 1/2. This is a relatively small change for this characteristic since in Figure 4, which shows the same measure in the Warm Springs Native American children, the total change is from 25 diopters down to 20 diopters, or a change of 5 diopters, compared with 2 1/2 diopters for the British subjects.

The primary reason for changing groups at this point, is to show the increase in lens power occurring at age 13 1/2 in the British population, which is similar to the increase in lens power occurring in the chimpanzee population age age 10 1/2. We believe that this increase in lens power portends the beginning of the development of myopia in any group, and therefore is an important indicator of the visual refractive changes taking place in any group of subjects.

Figure 4 presents similar results for the Warm Springs Native American children, for the males and females separately and combined. If one compares only the female group, one can see again a considerable decline in lens equivalent power with age, and a reversal of approximately 1 diopter occurring at

age 10 1/2 for the females, as well as a reversal of approximately 1 1/2 diopters occurring at age 11 1/2 for the males.
These results are in line with our earlier findings, since
females tend to develop myopia at an earlier age level than
males. However, if the results of the males and females are
combined, as they have been in the other Figures of the study,
this important change in the power of the lens, with respect
to the sexes, would be masked, as can be seen by examination
of the combined curve. It is necessary to keep sexes separate
in any analysis of visual refractive characteristics, since
these sex differences are of considerable magnitude in both
humans and apes, as well as monkeys.

Figure 5 presents the combined sex results for depth
of the vitreous chamber and total axial length. If the human
results are compared, it is clear that there is a great similarity between the shape and slope of the lines for the depth
of the vitreous chamber and the axial lengths. The actual
correlation between these two sets of figures for the Warm
Springs Native American subjects is approximately 0.95, thus,
generally speaking, myopia represents an increase in the depth
of the vitreous chamber.

If one returns for a moment to Figure 1, where there is
an increase in the depth of the anterior chamber, and the data
had been followed for a longer period of time, this increase
in the depth of the anterior chamber decreases, returning
approximately to the values it had prior to age 10 1/2, so that
the depth of the anterior chamber in older myopic children con-
tributes relatively little to the amount of myopia actually
developed. It is also clear that the curve for the chimpanzee
for both the axial length and depth of the vitreous chamber are
similar in slope, but it should be noted that the axial lengths
of the chimpanzees more closely approximate the axial lenghts
of the Native American children than does the depth of the
vitreous chamber approximate the depth of the vitreous chamber
of the Native American children. The average difference in
axial lengths is just over one millimeter, while the average
difference in the depth of the vitreous chamber is one and
three-quarters millimeter. This difference is accounted for
by the average depth of the anterior chambers shown in
Figure 2, where the chimpanzee has an average depth of the
anterior chamber of approximately 0.5 millimeters deeper than
the human subjects.

As can be seen, there are basic similarities in the
development of the visual optical characteristics of chimpanzees and humans, and to a lesser degree of certainity, because
of smaller samples, among gorillas and orangutans. These basic
similarities as well as defined differences, permit a high level

of generalization between the chimpanzee and the human in terms
of visual characteristics.

REFERENCES

   Albee, J. L. and Young, F. A., An evaluation of the
retinal-neural fading phenomena and its possible role in reading
difficulties. Unpublished Master's Thesis, Washington State
University, Pullman, Washington, 1976.
   Cowey, A., The basis of a method of perimetry with
monkeys. Quart. J. exp. Psychol., 15: 81-90, 1963.
   Cowey, A. and Weiskrantz, L., A perimetric study of
visual defects in monkeys. Quart. J. exp. Psychol., 15: 91-115,
1963.
   Moses, R. A., Adler's Physiology of the Eye, 5th Ed.,
St. Louis, C. V. Mosby Company, 1970, p. 438.
   Polyak, S. M. The vertebrate visual system. Chicago,
University of Chicago Press, 1957.
   Reynolds, V. The Apes, New York, E. P. Dutton & Co.,
1967, p. 74.
   Riggs, L. A., Ratliff, F., Cornsweet, J. C. and
Cornsweet, T. N.,The disappearance of steadily fixated visual
test objects. J. Opt. Soc. Amer., 43:495-501, 1953.
   Young, F. A. Visual refractive characteristics and the
subhuman primate. In Bourne, G. H.(Ed) Nonhuman Primate and
Medical Research. New York, Academic Press, Inc., 1973,
p. 353-379.

IMMUNOLOGY AND MELANOMA

H. F. Seigler

Duke University Medical Center
Dept. of Surgery and Immunology
Durham, North Carolina

One of the most feared and difficult cancers is the
malignant mole--melanoma. This malignant disease may be
highly invasive and progress rapidly to death of the patient,
may exist in harmony with the patient for a number of years
or may even spontaneously disappear. This variable clinical
course suggests that the host response to the tumor is an
important consideration. Numerous animal and human experiments suggest that melanoma is a very antigenic tumor and is
thus, potentially susceptible to elimination by the host
immune response. It is not too uncommon for the physician to
note that the primary tumor undergoes regression while metastatic deposits in the lymph nodes either persist or enlarge.
Also suggestive of the host control mechanism is the appearance of metastatic tumor long after the primary tumor has been
"completely" removed. A number of well documented cases describing the regression of widespread metastatic tumor have
been reported.

The destruction of a tumor by the host immune defense
mechanism is dependent upon an interaction between elements of
the immune response and tumor associated antigens (TAA) on the
tumor cell membrane. A number of investigators as well as
ourselves have provided serological evidence for a cell membrane antigen common to the malignant melanoma cell. The
evidence would suggest that the membrane TAA of different
melanomas, as detected by the sera of melanoma patients,
possess some degree of reactivity, although there is still some
question as to whether or not each melanoma may possess individually unique TAA in addition to those cross-reacting with the
TAA of other melanomas.

Interest in the use of hetero-antisera for the study of
melanoma TAA has arisen because of two problems encountered in

the use of melanoma patient sera for this purpose. When the
cell membrane melanoma TAA has been studied, it has been found
that no single serum from any melanoma patient will react with
all target melanomas. Secondly, the titer of such sera is
usually quite low. Hetero-antisera have the advantage of
being higher titered and of broader reactivity than patient
sera. The major problem encountered with the use of hetero-
antisera is concerned with the specificity of the resultant
antibody generated. In addition to responding against TAA,
immunized animals may also respond against human species,
organ, allo- and iso- antigens and it is only after antibodies
reactive with these normal human antigens are absorbed out,
that the anti-melanoma TAA activity of such antisera may be
realistically evaluated. In an effort to obviate their response against normal human antigens, we have utilized the
chimpanzee for production of anti-melanoma antisera. Because
of their phylogenic proximity to man, chimpanzees share many
antigens which are cross-reactive with humans. Because of
this, the primate immune response against many normal human
antigens is much weaker than that of more commonly employed
species, thus permitting a stronger, more specific anti-TAA
response. Tumor cells from a single patient have been utilized for immunization of a histocompatible selected chimpanzee.
The antiserum raised reacted against cell membrane TAA found
on all human melanoma cells tested, but not with tumor cells
from other spontaneous human tumor cell types. In addition,
the antiserum reacted with cell membrane antigens of human
fetal fibroblasts. Absorption of the antiserum with cells
from human fetuses reduced, but did not remove all the anti-
melanoma activity. This did not appear to reflect quantitative differences in antigen expression on melanoma and fetal
cells an additional absorptions of the antiserum of fetal
cells failed to further reduce the anti-melanoma titer of the
antiserum. Absorption of the antiserum with cells from any of
ten different melanomas removed all detectable reactivity
against both melanoma and fetal cells. These results may be
interpreted to indicate that at least two distinct TAA are
expressed on the melanoma cell membrane. One of these is
shared with fetal fibroblasts and the other apparently is unique to melanoma cells.

Using this chimpanzee anti-human melanoma antibody, we
have been able to demonstrate that melanoma cell membrane TAA
are released from the cell membrane in an antigenically active
form spontaneously as a result of membrane regeneration. We
have also found that these TAA are susceptible to proteolytic
digestion with pronase and trypsin, as a treatment of melanoma
cells with these enzymes renders the cells refractive to kill
by the antibody. A number of other investigators have suggested that the host immune response to the tumor was blocked

by either free tumor antigen that was released and combined
with either circulating antibody or cell bound antibody to
render the host response either its tumor ineffective. The
high titered chimpanzee anti-melanoma antibody has allowed us
to detect and characterize both free antigen and antigen-
antibody complexes in serum samples obtained from patients
who either have been successfully surgically treated for this
malignant disorder or who presently continue to demonstrate
metastatic lesions. Because of the specificity usually realized with antigen-antibody interactions, there has long been
interest in using tumor specific antibody to concentrate anti-
tumor agents on tumor cells. In doing so, many of the harmful
side effects of such chemotherapeutic drugs might be avoided.
Past studies have usually not been successful because the
antibody has either lacked suitable specificity for the tumor
associated antigen, or that the binding affinity was low. One
recent study, however, had rather dramatic success and has
stimulated a great interest again in this area. The antibody
was conjugated with an alkyating chemotherapeutic agent and
administered to a melanoma patient with disseminated tumor.
Following this treatment, the tumor systemically regressed.
We have utilized radio-labelled chimpanzee anti-human melanoma antibody in an effort to distinguish the specificity of
the antibody and its ability to localize to the harbored tumor.
Both immunofluorescent studies and radio-autography studies of
the removed tumor tissue have demonstrated that the antibody
did, indeed, fix specifically to the tumor cells. We are
presently conjugating the antibody with Technesium 99M in an
effort to develop a specific radioscanning technique for
demonstration of metastatic disease by this more sensitive
and specific technique.

      Although the diagnosis of melanoma is usually not diffi-
cult for the pathologist, approximately 10% of the cases of
metastatic disease can be interpreted as undifferentiated
tumor campatible with, but not diagnostic of melanoma. Because
current success with certain chemotherapeutic agents and
immunotherapeutic regimens demand specific diagnosis of the
tumor type, a more precise method for the pathologist is of
paramount importance. We have been successful in specific
serodiagnosis of either suspended tumor cells or reactions
with tissue slices utilizing the chimpanzee anti-melanoma
antisera. The serological results have been absolutely specific and can be done with little expense and relative ease.
This observation has opened the door for future development
by clinical pathologist of serodiagnosis for malignant diseases. Utilizing immunofluorescent techniques, the fluores-
cein conjugated antibody can be specifically utilized on
either tissue slices or done on frozen section material for
immediate diagnosis. The chimpanzee antibody has thus demon-

strated an immediate and quite useful tool for the clinician.

We are presently developing a radio-immunoassay for melanoma using the chimpanzee anti-melanoma antisera and isolated, purified melanoma tumor antigen. Similar radio-immunoassays have proven of great value in other disease states. Such an assay for insulin, renin, gastrin, and hepatitis antigen only to mention a few, indicates the potential contribution that such an assay could offer to the cancer specialist.

A more basic problem that we are investigating, concerns the biochemical nature of the human tumor antigen and the host response to these antigens. To say that tumor antigens appear to be altered fetal or histocompatibility antigens only says that they are transiently expressed normally during the gestational period. Nothing has been documented about the possible role or function that these antigens perform within the cell. One possibility that cannot be ignored, is that TAA are merely host components which have been somehow modified such that the host immune response recognizes them as foreign. One approach to solve this problem is to isolate the melanoma TAA for study apart from other molecular species. The chimpanzee anti-melanoma antibody has allowed us to initiate studies in this area. We have been successful in separating the antigen from the tumor cell membrane and can detect the antigen fraction obtained from chromatography purification. An immune absorbence column utilizing the specific anti-tumor antibody has permitted further isolation and purification. Further characterization of the antigen will aid not only in the understanding of the etiology of the transformation of the cell from a normal cell to a malignant cell, but may indeed, allow us to understand the functional role that these antigens play. The critical points concerning the immune mechanisms involved in the host tumor relationship may begin to be answered as specific antibody reagents are developed thus, permitting careful laboratory study. The importance of biomedical research in human cancer is more evident today than ever before, the obvious important role that the subhuman primate plays in this continued research is evident. Research data accumulated using this experimental animal, so close to man, has in the past and will continue in the future to be directly applicable to the human situation and thus, permits vital investigation that for moral and ethical reasons could have never been considered using human volunteers.

FIELD STUDY OF PRIMATE BEHAVIOR[1]

S. L. Washburn

University of California
Berkeley, California

In 1929 Yerkes and Yerkes published a major review of
the behavior of the great apes. They concluded that it was
"ludicrous" (p. 590) that scientists should rely on accident
and folklore for their knowledge of the behavior of the apes
and monkeys. Yerkes not only pioneered in the use of the apes,
particularly the chimpanzee, in psychological research, but he
also promoted the first field studies of ape behavior. The
monographs on the chimpanzee (Nissen, 1931) and the gorilla
(Bingham, 1932) mark the beginning of a new era. But the conditions in Africa proved exceedingly difficult for individuals
with no experience in fieldwork. It was Carpenter's study of
the howler monkey (1934) which showed what could be done and
which fully justified Yerkes's hopes and expectations. However,
in spite of Carpenter's success, there was little interest in
the field studies of that time. At Harvard the late E. A.
Hooton became interested, talked enthusiastically of fieldwork,
and dedicated his book, "Man's Poor Relations" (1942) "To
Robert Mearns Yerkes, student of primates, teacher of primatologists."

Harold J. Coolidge, Jr., who spoke to us yesterday,
organized an expedition to Asia for the Harvard Museum of Comparative Zoology. The principal purpose of the 1937 expedition
was to study gibbons -- Carpenter studied behavior, publishing
the results in 1940, and Schultz made anatomical studies which
appeared in numerous publications over a period of many years

[1] My thanks to colleagues and students here at Berkeley
for discussions on issues raised in this paper, to Mrs. Alice
Davis for editorial assistance, and to the L.S.B. Leakey
Foundation for financial support.

(summarized in Schultz 1969). But even these highly successful efforts did little to stimulate interest in primatology, and it was many years before Harold Coolidge's prophetic visions of a primate center became a reality. Even as late as 1955, in a paper published posthumously by James Gavan, Hooton complained of the lack of interest in behavior in general, and of Carpenter's studies in particular.

Today it seems surprising that there was so little interest in field studies of behavior, but it should be remembered that the situation in social science was little better. In the traditional studies of human evolution the data were equally unreliable. Hobbes, Locke, Montesquieu, and Rousseau relied heavily on traveler's tales, and voyagers' accounts were the source of a great deal of the data for the early social thinkers (Penniman 1965). In "The History of Human Marriage" Westermack (1891) described the gorilla family, and it did not occur to writers at that time that extensive fieldwork would be necessary before anyone could learn the facts of gorilla behavior. My purpose is not to review theories of human evolution, but to emphasize that all the theories rested on data which had been collected in a haphazard manner by untrained people.

Whether human behavior or the behavior of other primates was under discussion, the evolutionary sequences were based on misinformation. It was these hypothetical sequences which brought the theories of social evolution into disrepute, and it has taken many years to eradicate the misunderstandings caused by nineteenth-century evolutionism. We primatologists must remember that the reason many social scientists oppose evolutionary thinking is that it was necessary to remove the rubbish, and it has been by no means clear where the rubbish leaves off and useful biology begins.

Whether we are discussing social science or studies of primate behavior, the essential change from the old to the new comes with fieldwork. It is the difference between sitting at home and using the data Yerkes branded as "ludicrous" or going into the field and studying the animals under natural conditions. In this sense the fieldwork promoted by Yerkes, and first successfully completed by Carpenter, bears the same relation to the earlier supposed facts as the fieldwork of Malinowski and Radcliffe-Brown does to the earlier anthropology. When the scientist goes to the field to gather his own data, there are two kinds of changes -- humanistic and scientific. Watching animals is fun. The animals are worthwhile in their own right. They are fascinating creatures, and I think that anyone who does not have this feeling of real interest and desire to understand the animals as worthy subjects had better stay home. In the field it is the animals that determine what

is going to happen and not the scientist. One may go out before sunrise to locate the group before they move from where
they have been sleeping, but if it is rainy, they are likely
to sit in comfort while the observer waits and waits and waits.
It is exceedingly important to see that the desire to see the
animals must be so strong that the inevitable problems of
field work seem trivial compared to the satisfactions. Trying
to see the animals under undisturbed conditions, trying to
understand their way of behaving, is a profoundly humanistic
experience. One learns to know individuals -- their pleasures,
problems, and, possibly, their injuries. Observation of the
lives of our closest relatives is a moving experience and a
pleasure, but it is no guide to accurate collection of data or
interpretation. The fun and fascination must be regulated by
a developing science of behavior. Only knowledge gained from
experiments can eliminate anthropomorphism, and this is why
fieldwork, no matter how successful, must be related to the
laboratory.

When we think of the pleasures of fieldwork today it is
important to remember that this is the world of tropical medicine and the airplane -- two factors probably more responsible
for the increase in field studies than any intellectual reasons.
Not so long ago every traveler in the tropics expected to be
sick. Moorehead's "The White Nile" gives a vivid account of
the way explorers were sick, how they were carried on litters,
and spent months moving a few miles. In those days porters
often deserted and were beaten half to death or shot. It was
an extraordinary way of life, and when Kruuk (1972) wrote, "I
had a marvelous time in Africa," he was referring to the world
of today. A few decades ago people did not have a wonderful
time. As Harold Coolidge mentioned yesterday, Carl Akeley died
in the gorilla country.

The great apes all live under conditions that make observation difficult, but no such problem accounts for the mis-
understanding of monkey behavior. Macaques and baboons and
langurs are easy to observe because they all live close to man
and were often kept as pets. But whether monkeys were regarded
as gods or nuisances, no systematic effort was made to understand their behavior -- not in the earlier cultures of India,
the Far East, Egypt, nor among tribal peoples. This was not
due to lack of interest in the animals -- every culture has
animal tales and many have animal gods. Hunting was a major
activity until very recent times, but this did not lead to reliable information on behavior. Interest in careful study of
behavior is a product of modern science and many of the results
of such studies run counter to the common sense of the people
who are living where the animals are. It isn't just being
near the animals that will make a study of primate behavior,

the science has to be brought to the animals.

Seeing animals in the great game reserves is a tremendous experience, both humanistically and scientifically, and we must remember that the primates that we study in the laboratory survived in competition with all these other animals. If we are thinking of the factors which determine primate evolution we cannot abstract the primates out as if they were some private group; we should see what the changes are and we should cooperate with people like Kruuk and Schaller and others who examine, or **try** to examine, this whole complex. The rich descriptions of what animals actually do are essential first steps. But the history of science shows description alone is a very weak tool and I would stress this because so many animal behavior studies tend to be comparative and then stop at that level. The contribution of ethology has been the combination of naturalistic observation with ingenious experiments, not the study of unmodified natural behavior. Primate field studies will surely move in the direction of greater control -- Hans Kummer is the leader in this regard.

At the present time there is the tendency to treat the comparative study of behavior as a rather distinct discipline and I think that is a profound mistake. There should be a constant interplay between the study of the behavior and the comprehension of the underlying biology. It is not a question of one or the other or which one first and which one second. For example, in the primates (and this has been mentioned before at the meeting) there has been a delay in maturation. It is increasingly delayed in the order of apes and man, and described in great detail by the late Adolph Schultz. This phenomenon has been interpreted as providing time for more learning, and this is the way it is treated in our schools. While this may be the case in part, maturation has also been delayed in such forms as elephants and rhino, and the rhino particularly haven't learned much, in spite of the lag.

Since large animals mature more slowly in general than closely related small ones, and size seems to be related to length of life, the reasons for the delay are complex. For example, the large forms seem to have much slower heart rates than the small forms. At least larger size greatly reduces birth potential, and longer periods of immaturity may be consequences of longer periods of play and learning. But maturation and learning are not related in any simple way. And, more importantly, neither has been analyzed very satisfactorily. If the interest is social behavior, then maturation might be most usefully described in terms of maturation of the brain and not in years. For example, early stages in human maturation are in utero in monkeys and apes.

If we can show that social behaviors are based on particular structures in the brain, then we should make our social comparisons on the basis of the maturity of those structures and not use months or years as the basis for comparison. Is there any evidence that an orang has a capacity to learn more than highly social ground-living monkeys? Here is the orang, a large-brained ape, and yet how much does the orang really know? What kind of learning is supposed to have taken place in the orang that would not take place in much smaller animals and in living in much more complex social situations?

Interestingly, while social behavior is usually treated behavioristically with little regard for biology, sexual behavior has been considered from both points of view consistently for many, many years, as was indicated earlier this morning. The pattern then of studying social behavior in its basic methodology is very different from the pattern of the way sexual behaviors have been studied. I believe that the model of the study of sexual behaviors as developed by Frank Beach, for one, is a much more useful model than the separation of social behaviors in dealing with them as if they were something very separate. Aggressive behaviors, as Dr. Bernstein has shown us, provide useful examples of the importance of considering field studies, experimental studies, and biology. The kinds of situations eliciting aggressive behaviors have been reviewed by David Hamburg and Jane Goodall in chimpanzees particularly, Mark and Irvin for man. In spite of the general similarity in basic biology, there are marked species differences in the anatomy of such behaviors as bluff, threat, and fighting.

The fieldworker sees these behaviors and their effects within the group, between groups of the same species, and between species, the point that Bernstein illustrated. In most instances the male appeared to be far more aggressive than the female. The anatomy of the females has been traditionally described as a percentage of the anatomy of the male and this is true even now. The female, they say, has 90% of the anatomy of the male -- a very peculiar way of describing anatomy. In free-ranging primates females sleep, sit, locomote, and eat in the same way as the males. The basic social system is that of the females according to many recent studies. What the males then add to the anatomy of the females is the anatomy of bluff and aggression (McCown 1977). When described in this way, and we deal with all these different structures in anatomy of the male as plusses, we find that they're much bigger than they appear to be than if we play this game that the female is a certain percentage of the male. For example, male baboons are about twice the size of females, but the fighting muscles, such as the big temporal muscles, are four times the size of that of the females. In the vervet monkey one of the adult

males has 45 grams of temporal muscle; the female has 15. If
you look at the animal, or if you consider the female as a
percentage of the male, these differences don't seem to be very
great, but when you measure them, they are.

What about chimpanzees? Why should male chimpanzees be
much bigger than the females? For a long time humans viewed
animals as violent, difficult, and aggressive -- Du Chaillu's
gorilla, for instance, bending the gun in his hand and leaving
tooth marks on the barrel -- you can't be much more violent
than that. Then we went into a period that we may be getting
out of (we're certainly still in it), when the desire to see
the animal in nature was to see a friendly, democratic, middle-
class American chimpanzee, and that's what we got. What happened to the chimps in the last few years? They didn't change
-- the conditions of observation, the nature of observation,
the way of dealing with the observations changed. So this
animal becomes aggressive, hierarchy-seeking, territorial,
fighting its own kind, eating a baby chimpanzee (stranger from
another group). This is not the chimpanzee many have described.
The same holds for gorilla. Transformation from this very
peaceful and inactive gorilla described by Schaller (1962) --
and this is not meant to be a criticism of Schaller; he had immense courage to do that study at the time that it was done --
but Fossey's gorillas pound each other, kill infants, in one
case ate infants (Fossey in press).

The point here is that we have to empathize with the
animals, we have to see the animals, we have to devote this
effort to the animals, but if we don't control the observations and have some experiments, some awful mistakes will be
made. And just as the social scientists have a right to protest what was said from an evolutionary point of view, social
scientists will also have a right to protest the present primate studies -- unless we put our own shop in order. You put
it in order partially by good field work, but much more by
careful experimentation and laboratory analysis. For example,
the statement that human females are distinct from all other
primates because they are continuously sexually receptive runs
all through the literature on human evolution. Further, it
has been repeatedly claimed that this has been a major factor
in the origin of the human family and distinctively human
social behaviors. Yet only this morning Dr. Nadler told us
that female orangs are also continuously receptive. So the
least social of the monkeys and apes has the physiological
condition supposedly causing human social behavior. One experiment has fundamentally changed all contemporary theories
of human behavioral evolution and of the relations of sexual
receptivity and society. This shows the critical importance
of experiments for our theories, especially as a background

for fieldwork.

I want to speak about the phylogenetic position of the great apes because this has been mentioned repeatedly. Behavior students have tended to, by and large, regard chimpanzees as very close to man. It has been advantageous for them to do so, if you want to look at it that way. On the whole, the great apes have been considered as closest, but separated by a very comfortable amount of time -- at least 20 million years. As late as 1972 an eminent paleontologist published his contention that man and chimpanzee were separated for more than 35 million years, more than half of the length of the age of mammals. To go back 35 million years, and then up 35 million years means the animals were separated by 70 million years of evolutionary time, and this was widely believed until recently.

We now have biochemical methods of determing the relationships between the animals which are quantitative, and which are the same no matter in which laboratory the determination is made. DNA hybridization, a technique which is being modified and improved was applied to primates eight or nine years ago. The sequence of amino acids and proteins is another way of determining relationship. The way of doing it fastest is through immunology. Here are a few figures from Vincent Sarich and John Cronin from Berkeley on the immunological comparison of the great apes and man. These are in immunological distance units. They represent differences of a molecular nature and they correlate .98 with the sequence of amino acid differences and with DNA hybridization.

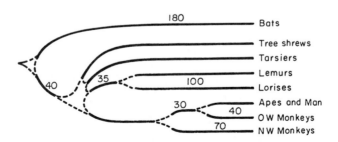

Fig. 1.   Albumin plus transferrin phylogeny of Primates. Modified from Sarich and Cronin by Michelle Freier.

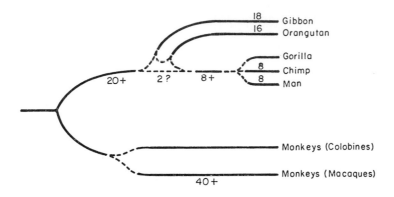

Fig. 2. Albumin plus transferrin phylogeny of Apes and
Man. Modified from Sarich and Cronin by Michelle Freier.

        Going from man away -- 8 units to the chimpanzee and the
gorilla, who are equally different from each other and both
equally different from man. This kind of thing, you see, lets
us put some real estimates of time and adaptation into the
picture. If gorilla, man and chimp are equally related, a
great deal more has happened in the human genotype than in the
others and this offers a chance of seeing when evolution goes
rapidly and when it doesn't. Orang -- 16 units out, twice as
far. Old World monkey -- more than 40 units out. Putting
this together, it means that the great apes and man had a long
period of common ancestry, whatever that may be in years. So
the similarities between man and chimpanzee are not due to
parallel or converget evolution, but to a long period of shared
ancestry. I think the proof of this is very important in the
way we look at the comparisons and the behaviors. New World
monkeys -- 70 units out. Prosimians -- 140 units. Then only
40 more units in common to the basic primates. Therefore the
primates as a group shared a very short period of time in
common and then divided into the different prosimian groups.
Old World and New World monkeys shared a very long period of
time in common before the New World monkeys presumably rafted
over from Africa, before South America had drifted that far
away from Africa. These 8 units of 140 constitute the difference then between man and chimpanzee and man and a prosimian --
8 units out of almost 200 -- the difference between man and
chimpanzee and man and a primitive primate. This, I think,
makes man and chimpanzee and man and gorilla far more closely
related than we ever thought they were from an anatomical

point of view. Everything that Dr. Yerkes and everything that
the Yerkes Lab stands for is supported in the strongest way by
these biochemical comparisons. We should start all our presentations with a little chart giving this information so that
we know the kind of distances that we're talking about.

In the future I think it's going to be possible to transfer those units into time -- and there are some jokers in that.
But with DNA, with sequence of amino acids in the proteins,
with the immunology and with other methods which are being
developed now, and with the different molecules evolving at
different rates, there is a whole matrix of information, and
that can theoretically be changed into time. The problem at
present is that there is no one point that anyone can agree on.
We have a scheme for getting around that, but we're just trying
it and I think it is going to be possible to translate molecular distance into time.

To get back to the point that Peter Marler made, and,
particularly, a point that Karl Pribram made, relative to biology, behavior, language, and learning. Chimpanzees cannot be
taught to speak. The language studies so elegantly reported
on yesterday are all designed to use other modalities. The
question then is, is that a continuum or is something new,
something different coming in? In Robinson's experiments (1976)
on implanted electrodes in rhesus monkeys, all the normal
sounds were produced by electrodes implanted in the limbic
system near the midline. Electrodes implanted in such a way
in a human being would not produce any language. In monkeys
(Myers 1976) the removal of substantial amounts of cortex does
not affect any sound and hardly affects facial expression at
all -- very surprising. So it appears that this is the point
Marler was making, like the degree of interrelation between
these two systems that the phonetic sound code system of man
is controlled primarily by the cortex. It is always paralleled
by the emotional system -- the two are working together, as so
well described before. Nevertheless, talking is different from
communicating with gestures. Our research should keep communication by gestures and communication by talking with a phonetic
code as separate systems. We should consider the fact that it
may well be that man's talking is just as unique as we used to
think language was. What the language studies show is that
cognitive similarity, the base on which the talking is added,
is very great. But using the word language for both the limbic gestural system and for the cortical phonetic code system
is misleading and confuses the interpretation of the experiment.

I'd like to make three final points on: visits to fieldwork, field experiments, and conservation. First of all, it
is impossible for most scientists who were using primates in

their research to do fieldwork. It is not practical to suggest
that everyone using primates in the laboratory should, but
primatologists working in laboratories might visit ongoing field
studies. This would not take much time. Those studying the behavior of captive animals (Alcatraz behavior as I like to call
it) would also see the animals under these very different conditions. I think of Dr. Menzel's work on the charting of how
the chimpanzees really went. In looking at primates in the
field they appear to be doing very complex things. You can
chart the way they go with a little cluster of lines indicating
that this is the way they went, that these are the reasons they
went, but the field situation is too complex. The beauty of
Dr. Menzel's experiments is that it's simplified to a point to
where you can see what's going on. Fieldwork will always be in
trouble because it's too complex; it has to be simplified. The
visits to the field would enrich both the fieldwork because the
experimentalist would be talking to the person who does the
fieldwork, making suggestions, and asking the awful questions
students are wont to. How many creatures are in the troops?
How many troops do you see? It puts anyone on the defensive
almost at once. The only animal easy to count is the orang.
Field visits are important; field experiments are important.
Rather than arguing as to whether these animals definitely eat
meat, go ahead and give them some and see if they eat it or
not. Give them a live hare or whatever it is you think they
eat. Speed up the process. If it's stated that langurs kill
infant langurs, introduce an infant langur and see what really
happens. These events occur so rarely, and you are so unlikely
to see them, that you've just got to do field experiments or
we're never going to get the data.

        Finally, consider conservation. The human population is
expanding at a colossal rate and regardless of what we may
think about the problems of protection, if these animals are
not protected, there aren't going to be field studies. It is
very naive to suppose that the poorest peoples in the world
should pay the cost of the parks. They aren't. And it is not
fair to ask them to do so. Through some international organization it should be possible for those of us who want the
animal reserves kept to pay a share. A reserve costs so much
per year in terms of it not being used for farmland or other
such purposes. Through the United Nations or some other
institution, the rich nations should take the responsibility
for the reserves rather than the poor.

        The problems of conservation and protection of the ani-
mals are but one part of the whole problem of population,
pollution, and environment. These world-wide problems are of
unprecedented magnitude and importance. I wonder what chimpanzees would say if they could perceive the problems. They

might think that their problems are rather simple compared to those of <u>Homo</u> <u>sapiens</u>. If they could speak, they might say, "Never send to know for whom the bell tolls for thee."

REFERENCES

Bingham, H. C. (1932). Gorillas in a native habitat. Carnegie Inst. Publication No. 426.

Carpenter, C. R. (1934). A field study of the behavior and social relations of the howling monkeys. <u>Comp</u>. <u>Psychol</u>. <u>Monogr</u>. <u>10</u>, 1-168.

Carpenter, C. R. (1940). A field study in Siam of the behavior and social relations of the gibbon. <u>Comp</u>. <u>Psychol</u>. <u>Monogr</u>. <u>16</u>, 1-212.

Goodall, and D. A. Hamburg (1975). Chimpanzee behaviour as a model for the behavior of early man. <u>In</u> Hamburg, D. A. and H. K. H. Brodie, eds., American Handbook of Psychiatry VI, 14-43.

Hamburg, D. A., McCown, E. R., and Goodall, J. eds. The Behaviour of the Great Apes. Perspectives on Human Evolution Series, Vol. 5.

Hooton, E. A. (1942). Man's Poor Relations. New York, Doubleday.

Hooton, E. A. (1955). The importance of primate studies in anthropology. In Gavan, J. A., ed., The Nonhuman Primates and Human Evolution. Detroit: Wayne University Press.

Kruuk, H. (1972). The Spotted Hyena. University of Chicago Press.

Mark, V. and F. R. Ervin (1970). Violence and the Brain. New York: Harper and Row.

McCown, E. R. (1977). Patterns of Sex Differences: Cranial Anatomy of Old World Monkeys. Doctoral dissertation, University of California, Berkeley.

Moorehead, A. (1960). The White Nile. New York: Harper and Row.

Myers, R. E. (1976). Comparative neurology of vocalization and speech: proof of a dichotomy. Annals of New York Academy of Sciences 280, 745-757.

Nissen, H. W. (1931). A field study of the chimpanzee. <u>Comp</u>. <u>Psychol</u>. <u>Monogr</u>. <u>8</u>, 1-122.

Penniman, T. K., ed. (1965). A Hundred Years of Anthropology. London: Duckworth.

Robinson, B. W. (1976). Limbic influences on human speech. Annals of New York Academy of Sciences 280, 761-771.

Schaller, G. B. (1963). The Mountain Gorilla: Ecology and Behavior. University of Chicago Press.

aggressive expression (Bernstein et. al., 1974c). Our perspective has also included ontogenetic studies, which are still
underway, and comparative studies using a variety of old world
monkey species and even a few new world monkeys and an occasional ape group. These comparative studies represent the only
technique we have used in considering evolutionary aspects of
the problem, but these techniques also allow for some cautious
extrapolations to our own species of primate.

     In our comparative studies, we quickly recognized broad
similarities and differences in the various aspects of aggression studied in each of the species. We noted that in several
taxa, unfamiliar adult male conspecific intruders were among
the most potent stimuli releasing the more intensive forms of
individual aggressive expression and extreme mobbing behavior
by groups. This directed our attention once again to the
immediate causes of aggressive behavior in primates. Scientists rarely talk of "cause" in the strict sense, leaving discussions of causation to philosophers, but correlations may
imply a casual relationship, and we therefore began to examine
the stimulus characteristics of subjects eliciting aggression.
The age and sex attributes had proven critical, so it was not
unfamiliarity alone that accounted for the hostile reception
intruder males received. The unfamiliar appearance of the intruder did seem important and yet we felt that being an
individual with many familiar attributes, but without a shared
social history, was the critical aspect triggering group re-
jection of unfamiliar conspecific adult males. We therefore
determined to examine the degree of unfamiliarity of the intruder's appearance. We knew that markedly different animals
such as; birds, lizards, insects or even other large mammals;
were not instantaneous triggers of attack. We suspected that
socially living primates would somehow maintain monospecific
sociality without the need for violent responses to extraspecifics, and yet we realized that monkeys must recognize the
similar appearance of monkeys of related taxa. We therefore
instituted a series of tests using extraspecific adult male
intruders to see if they were as vigorously attacked as conspecifics and to search for alternative mechanisms which
operate to maintain biological species specific sociality.

     The new test methods paralleled the original intruder
intruduction tests, but now each intruder was a member of another taxa, some being cogeneric forms and some more distally
related. Both individual and group tests were scheduled, but
inasmuch as we could more readily control single animal introductions, they were always run first. Further, inasmuch as
conspecific introductions had demonstrated the variability of
aggressive expression among primate taxa, we ran each extraspecific introduction with counterbalanced controls. This also

served to control for relative sizes, just in case our previous data on age parameters had been confounded by the correlation between age and size.

Inasmuch as macaques have the reputation for being among
the most aggressive of primates, we began with: the old laboratory standby, the all too familiar and nearly intractable
rhesus monkey (Macaca mulatta); the very large pigtail macaques
(M. nemestrina) and; the much smaller crabeater macaques (M.
fascicularis). Our comparative perspective convinced us of
the necessity of extra generic controls, and mangabey (Cerco-
Cebus atys), guenon (Cercopithecus mitis), vervet (C. aethiops),
and new world monkey (Cebus albifrons) controls were included
in the series, as well as another macaque group, the bonnets
(M. radiata) selected to match the size and composition of our
vervet group.

The series began with counterbalanced single animal introductions, including five pigtail males and five rhesus
males each individually introduced to a host of the other species, and also into the crabeater host group. We also tested
single representatives of bonnets, guenons and vervets with
the crabeater group and tried guenons and capuchins with the
vervet group. Conspecific controls, as well as a female control, were inserted into the series, and no animal served as
both subject and host. To verify the generality of a group's
responses we replicated one pigtail introduction using a
second rhesus host group.

In all of these tests we never used the same host group
more than once a week, and in the cases where a single subject
was introduced to two different groups, we always scheduled at
least a two week rest period between introductions.

At the conclusion of these single animal introductions,
we felt emboldened enough to try group to group mergers. We
first introduced the pigtail and crabeater host groups (approximately 40 and 55 animals respectively) to each other by
opening a connecting door between their living quarters and
leaving it open for the next 30 days. Subsequently, we introduced the vervet group to the bonnet group (approximately 10
and 12 animals, respectively) and later the same vervets to a
group of nearly 40 mangabeys.

In our previous studies of conspecific introductions, we
had seen an initial period of immediate and intense aggression
directed to the intruders, led largely by residents of the
same age-sex class. This initial period of attack was followed by a rapid reduction in the frequency and intensity of
agonistic behavior and the slower establishment of positive
patterns of social interaction. Individual intruders had been
mobbed and rapidly defeated by the group. They then went
through a period during which individual dominance relation-

ships and alliances were established and , finally, were fully
integrated into a unified social organization. In conspecific
group mergers, the sequence was much the same, beginning with
organized group to group fighting until one group defeated the
other. Members of the defeated group were then integrated
individually into the victorious group, but without regard to
their previous relationships in the now defunct organization
of the defeated group. The social structure of the victorious
group remained intact with the new members being added onto
the basic structure.

     If we had began our series of extraspecific introductions
believing that the primary function of aggression is to exclude
unfamiliar animals from a social group, then we would have predicted that extraspecifics would elicit at least as much aggression as unfamiliar conspecifics. If, on the other hand we
had believed that aggression functions primarily to establish
social relationships among conspecifics, we would have predicted that extraspecifics would elicit little aggression.

     In fact, neither prediction alone adequately accounts
for our findings. Extraspecific introductions did not produce
an immediate release of agonistic behavior, not did they preclude later episodes of serious fighting. Initial responses
included high frequencies of sniffing, touching and slow approaches, suggesting that residents were more curious than
immediately hostile. Consequent interactions were not readily
predictable. Extraspecifics were not attacked immediately, nor
were they accepted into the existing social units. The lack of
initial hostilities did not indicate an advanced state of individual acceptance, because the aggression that did take place
included the most violent forms of aggression, ordinarily not
seen in established groups of conspecifics. Moreover, the
usual attenuation of attack contingent upon the receipt of
proper submissive signals seemed to be abrogated by the extraspecific attributes of the target; defeated extraspecifics
emitting the full repertoire of extreme submissive signals
were nonetheless subjected to violent physical assault, rather
than more token forms usually directed at conspecifics in
similar circumstances. Even after a defeat, extraspecifics
were not immune to further severe punishment, and repeated
agonistic episodes failed to integrate them into the social
unit.

     Extraspecifics animals were, therefore, neither subject
to the same immediate attack as unfamiliar conspecifics, nor
integrated into the existing social group. The initial level
of agonistic responses may have been lower than that experienced by conspecific intruders, but these levels failed to decline to the low levels seen after conspecifics are fully integrated into a group. More importantly, the quality of attacks

did not change rapidly from damaging forms of aggression to
more token and less dangerous forms of expression. In cross
species group mergers and in mixed taxa group formations elsewhere reported, (Bernstein and Cowlan, in press), we found
persistent prolonged periods of serious wounding throughout the
studies.

Considering the group to group mergers as a single extreme
and dramatic test, first let us consider the merger of the two
macaque groups. Within the first few minutes after the door
between the two compounds had been opened, members of both
groups entered the opposite living areas and the animals intermingled throughout all of the available space. No serious
fighting was observed in the first hour and no group to group
confrontations developed. Over the next 30 days, however, we
recorded a pattern of persistent low frequency aggression which
included damaging forms of contact aggression. The much
smaller crabeaters generally came out a poor second best, and
although they became more cautious, their group neither dis-
solved nor was defeated by the pigtails. Instead, the crabeaters became more cohesive and continued to resist the pigtails in each agonistic episode using the full range of crabeater aggressive responses, rather than restricting their
responses to submissive signals characteristic of subordinate
macaques when provoked by higher ranking animals with established dominance relationships.

Agonistic episodes were brief, and explosive. Group to
group fighting never developed, although a few close associates of an individual, or nearby adult males often intervened.
Several crabeaters received serious injuries, and some might
have been fatal if not for prompt and effective veterinary
treatment. A list of treated wounds, however is a misleading
indicator of relative relationships inasmuch as many pigtails
also required medical treatment, but almost always because of
conspecific fighting induced by repeated captures and handling.
The pigtails not only wounded crabeaters more often than vice
versa, but also wounded one another more often when confined
for captures necessary to remove injured animals for treatment.
Wounding patterns therefore reflected more the readiness and
ability of the two species to slash, than a dominance relationship between species.

At the end of 30 days we had enough data to demonstrate
that no basic changes in social organization had transpired or
were imminent, and that no accomodation to the other group
could be evidenced by changes in the intensity or frequency of
agonistic encounters. Many forms of positive social exchanges
did emerge, with some individuals participating in cross taxa
grooming, play and even sexual responses, but the basic
organizations of the two groups remained intact.

The vervet group was introduced to the bonnet group to test the hypothesis that extrageneric forms might show a different pattern, but little new information was obtained by the second day, when the study was terminated after the first serious injury. Again, neither group was defeated, neither adjusted to the other, and an unacceptably high level of contact forms of aggression persisted from start to termination.

We hypothesized that much of the failure was related to some attributes which were common to the genus Macaca and that the behavior dispositions of other primate genera might not preclude living in harmony with another species in the same living area. The same vervet group was therefore scheduled for introduction to the much larger mangabey group. Although the mangabeys had the same numbers and physical size of the larger macaque groups, and the physical potential to inflict serious injuries, we believed they lacked the behavioral dispositions of macaques. Fortunately, we were correct and neither group showed the propensity to bite and slash one another, which the macaques had so readily displayed. Otherwise the merger was not different than those involving macaques. The two groups continued to maintain independent social structures with a low but persistent level of agonistic encounters. The quality of these exchanges has not varied significantly from the first hours of introduction, i.e. aggressive expression has not shifted to more token responses, it has simply seldom included the more damaging forms. A low level of affiliative interactions were also recorded. The only real contrast with the pigtail-crabeater merger has to do with the lower frequency of damaging aggression such that it is possible for these nonmacaques to share the same living space without an unacceptable risk of serious injury.

We have therefore concluded that the stimulus properties of an individual includes species specific signals which influence the nature of social interactions. Conspecifics of particular age, sex and behavioral histories may be subjected to immediate aggressive response by an unfamiliar group, but these responses are directed at repelling the intruder. If the intruder is defeated and cannot escape, there is a shift to more token forms of aggression which integrate the new animal into the existing social order. Once accepted into the group, an individual suffers minimal additional injury as social mechanisms exist which regulate aggression.

An extraspecific, on the other hand, does not elicit immediate attack, but anytime he does come into conflict with group members, he will be subjected to the full range of aggressive and defensive responses. Not being integrated into the group, social mechanisms limiting the expression of aggression between group members do not apply. Although the extraspecific

does not suffer an immediate initial attack, he will suffer
periodic attack into the indefinite future as he comes into
conflict with group members. Extraspecifics have neither the
stimulus properties which elicit group mobbing to repel conspecific strangers, nor the stimulus properties which allow
for the integration of conspecifics into a group resulting in
a shift to token agonistic exchanges which reaffirm relationships once established.

I think it important in a paper of this sort on aggression, involving the experimental induction of agonistic responses, to report the price of the study to the subjects used.
I am happy to state that in this series, not a single animal
died as a direct result of injuries received while participating in these studies, and that injury and suffering were
minimized by prompt and efficient veterinary treatment effec-
ted by a close working relationship between investigators and
veterinary staff.

REFERENCES

Bernstein, I. S.   1970. In "Primate Behavior". L. A. Rosen-
     blum (Ed.), vol. 1, Academic Press, New York. 71-109.
Bernstein, I. S.   1971. In "Behavior of Non-Human Primates".
     A. M. Schrier and F. Stollnitz (Eds.), vol. 3, Academic
     Press, New York, 69-106.
Bernstein, I. S. 1976. J. Theoret. Biol. 60, 459-472.
Bernstein, I. S. and Gordon, T. P. 1974. Amer. Scientist,
     62, 304-311.
Bernstein, I. S. and Gordon, T. P. In Press. In "Macaque
     Systematics" (title tentative). D.G. Lindburg (Ed.),
     Plenum Press.
Bernstein, I. S., Gordon, T. P. and Rose, R. M. 1974a. In
     "Primate Aggression Territoriality and Xenophobia",
     R. Holloway, (Ed.) Academic Press, Inc. New York,
     211-240.
Bernstein, I. S., Gordon, T. P. and Rose R. M. 1974b.
     Folia. Primatol. 21, 81-107.
Bernstein, I. S., Rose, R. M., and Gordon, T. P. 1974c.
     J. Human Evol. 3, 517-526.
Kummer, H. 1971. "Primate Societies". Aldine Atherton,
     Chicago, p. 160.
Rose, R. M., Bernstein, I. S. and Holaday, J. W. 1971.
     Nature, 231. 366-368.
Rose, R. M., Bernstein, I. S., Gordon, T. P., and Catlin, S.F.
     1974. In "Primate Aggression Territoriality and Xeno-
     phobia". R. Holloway (Ed.) Academic Press, New York,
     275-304.

Rose, R. M., Bernstein, I. S. and Gordon, T. P. 1975.
    Psychosomatic Med., 37, 50-61.

A REVIEW OF CROSS-MODAL PERCEPTUAL RESEARCH
AT THE YERKES REGIONAL PRIMATE RESEARCH CENTER

R. K. Davenport

School of Psychology, Georgia Institute of Technology
and Yerkes Regional Primate Research Center
Emory University, Atlanta, Georgia

K. J. Pralinsky

Yerkes Regional Primate Research Center
Emory University, Atlanta, Georgia

I wish to express my appreciation for being asked to participate in the Yerkes Centennial Conference and to recognize the importance of Robert M. Yerkes, the man and the scientist, in shaping the careers of most of us here and in strongly and permanently influencing the course of the disciplines with which he, and now we, are associated.

Many of the ideas and papers presented during this meeting reflect a changing Zeitgeist of American comparative psychology, animal behavior and anthropology. On the one hand, there is a continued chipping away at the strict dichotomy between Homo sapiens and other animals which has been based, for example, on the use of tools, the transmission of acquired information across generations (culture), and language. On the other hand, it appears that investigators are asking different kinds of questions and providing alternative explanations of primate (and other animal) behavior. This change, or perhaps just a resurgence of a tradition supported by Robert M. Yerkes, has come about to a great degree because of work with the great apes, whose behavioral complexity has become increasingly obvious in recent years, especially from the language-related

This work was supported in part by a grant from the National Science Foundation and by Grant RR-00165 from the National Institutes of Health, Division of Research Resources.

investigations. An important result of this (which is also occurring more broadly in general psychology) is a recognition of the awkwardness of conceptualization and the explanation of some phenomena without the use of cognitive concepts. These, of course, may be amenable in time to reductionist analysis; however, at this point, a more cognitive treatment of certain issues has considerable advantages.

I would like to describe a behavioral research program conducted over the last several years at the Yerkes Regional Primate Research Center which has import for neurological issues and underscores the cognitive approach just mentioned.

The sensory modalities of most human adults operate in concert as an integrated system of systems. Information received via one modality is coordinated in some manner with information from other modalities. Thus, a person can recognize an object seen as being the same (or at least very similar) to another object, in all respects physically alike, when access to the second object is through touch alone. Although the eye and the hand are receptive to different kinds of energy, a judgment can be made as to whether the objects are the same, different, or similar, even with marked discrepancies in the information originating in the different sensory modalities. Thus, a tool which is only seen can be judged the same as another which can only be felt. This is such a commonplace occurrence with familiar objects that we are seldom aware of this extraordinary perceptual ability which extends even to objects previously unknown. Presumably, much adaptive behavior depends on this intermodality integration; however, the extent to which it is shared with psychologically and neurologically unusual humans, infants and other animals is not clear. Indeed, the phenomenon is not well understood in normal human adults despite its importance as a scientific and philosophical issue regarding nativism, empiricism and the unity of the senses, extending back to the early 1700's with Locke and Berkeley. von Senden's (1932) work on early blinded people whose sight was restored in adulthood provided some interesting observations, but it has been only in the last ten years or so that systematic experimental research aimed directly at cross-modal perception, especially in nonhuman animals has received much attention.

Cross-modal or intermodal perception is a general term to describe, on a psychological level, the interrelatedness of the senses. Although there are several operations for its study, two major methods for assessing similarities and differences of complex stimuli have been used; cross-modal transfer and cross-modal matching-to-sample. Before continuing, a word on the term 'cross-modal transfer' is in order. As used by some investigators, the term 'transfer' seems to apply to some sort of interchange of information between two or more modalities occurring in the central nervous system. As used by others, it refers to a method of study, the transfer of training procedure, well known to psychologists. In our present ignorance of an interchange mechanism, there seems little value in using the term in its first sense and, to reduce confusion, I suggest that it be limited to its more conventional operational-procedural definition. Transfer of training refers to the influence of learning one task on the learning of a second task. If the original learning facilitates the later learning or performance, positive transfer has occurred; if later learning is more difficult, then negative transfer has occurred. Used to study cross-modal perception, for example, the subject learns to discriminate a cube and a sphere first by touch and is then presented visually the cube-sphere discrimination. If the subject learns the visual discrimination better than would have occurred without the first experience, then cross-modal transfer of the learned discrimination is said to have occurred and cross-modal perception is presumed. There are several limitations with this method. Since subjects usually receive multiple problems and trials, good performance in the second modality may be attributable to some general facilitation due to the initial learning or to rapid learning in the second modality unrelated to the specific experience in the first modality. Additionally, because of the time delay and change of modalities, the subject may treat the problem in the two modalities as unrelated tasks, with the result that no positive transfer (or even negative transfer) occurs.

The matching-to-sample procedure, the second approach, allows simultaneous or delayed comparison of the discriminanda (or percepts) in the two modalities. It requires the subject to match an object presented in one modality with an object (often one of several) presented in another modality. A problem is given with a set of discriminanda repeated each trial. This is followed by numerous other problems until a clear criterion of accuracy has been reached. Here, as in the transfer paradigm, since discriminanda are repeated, the subject may be learning the association of a sample and its match by rote. After training to criterion, it seems certain that the subject is either: 1) associating the two objects by rote; 2) performing a rapidly learned conditional discrimination; or 3) matching-to-sample on the basis of a concept or strategy of cross-modal stimulus equivalence. In testing these alternatives, subjects may be given "unique" matching-to-sample problems, each composed of new objects never encountered before. Only one trial is given on each problem and performance is analyzed only on those single trials. (In our own work, 40-60 unique trials have routinely been used.) In this case, since there is no opportunity for specific learning, correct matching occurs by chance or the subject is responding on the basis of the perception of cross-modal stimulus equivalence. The same end can be reached in the transfer design if analysis is limited only to the first trial of the problems presented in the second modality.

Indeed, although different in procedure, the two methods are conceptually the same. In the transfer procedure, the subject, by trial and error in the initial discrimination, must discover the object which is to be chosen when the same problem subsequently is presented to the other modality. This "correct" object is the equivalent of the sample in the matching-to-sample procedure.

If facilitation is the object of experimental interest, then the transfer of training method with repeated trials in the second modality is a reasonable procedure; however, if perceptual equivalence, i.e. if A in the eye then A in the hand, is the issue, then a matching-to-sample with unique trial procedure is the **more** powerful method.

Although auditory-visual cross-modal transfer of an intensity or pulse train discrimination has been found in several infra-primates (Over and Mackintosh, 1969; Ward, Yehle and Doerflein, 1970; Yehle and Ward, 1969), it was believed that visual-haptic cross-modal perception of complex patterns was limited to human beings and was perhaps mediated by language. (The term 'haptic' is used instead of 'touch' to underscore the active, palpating, information-seeking character of the hand in exploring objects.) The contention for man's pre-eminence in visual-haptic cross-modal perception was based on negative evidence. First, in the brain of the nonhuman primate, the auditory, visual and somesthetic association areas are relatively independent in that there is a paucity of cortical-cortical connections among these areas which, presumably, would make impossible or impede cross-modal functioning (Geschwind, 1965; Lancaster, 1968). Second, attempts to demonstrate haptic-visual cross-modal perception in non-humans met with repeated failures, or at least great difficulty (Ettlinger, 1967; Ettlinger and Blakemore, 1967, 1969; Wegener, 1965; Wilson and Shaffer, 1963), whereas it has been shown frequently in normal human adults. To explain this difference, language was invoked as a mediator to bridge the modalities (Ettlinger, 1967) since some studies had shown an increase in cross-modal abilities with improvement of language in human children (Blank and Bridger, 1964).

Experiments performed since 1970 have drastically altered the views expressed above. Our own research at the Yerkes Center has clearly shown that apes have the capacity for visual-haptic cross-modal perceptual equivalence of multidimensional objects or representation thereof. Subsequently, monkeys were shown to have similar abilities (Cowey and Weiskrantz, 1975; Weiskrantz and Cowey, 1975). Preverbal human infants (below one year) have also been found to have a visual-haptic cross-modal perception without specific training (Bryant, Jones, Claxton and Perkins, 1972). Thus, the basis for speculations that neuroanatomical characteristics limit the capacity of non-humans for intermodal perception requires revision and the verbal mediation hypothesis, insofar as it proposed that language is essential for cross-modal perception, has been clearly disproved.

An important key in the chain was our demonstration (Davenport and Rogers, 1970) of haptic-visual perception of cross-modal equivalence in apes using a matching-to-sample procedure and complex stereometric objects. This is not only put us on the trail of other cross-modal phenomena, but in addition, seemed to restimulate other investigators in the United States and Great Britain to further pursuit of cross-modal work in nonhuman primates. Our procedures employed an apparatus for displaying a visual (or haptic) sample, together with the appropriate choice objects. For example, the subject saw a sample and felt two other **objects**, one of which was a match. A slight tug on the match produced a reward. After considerable training, the "unique" trial test problems were used and the phenomenon confirmed. From the animal's point of view, operation of the apparatus was rather complex in that a number of tasks had to be learned, e.g. looking at the visible sample, reaching into an aperture, feeling objects, selecting a match, and so on. The success of the apes in accomplishing this task emphasizes their importance as experimental subjects, especially in the early phases of some research, when the most appropriate apparatus and most economical training procedures are unknown.

In essence, the behavioral characteristics of the ape allow the experimenter greater latitude in apparatus design and training procedures. Monkeys failed completely with this procedure and apparatus, not because of the absence of cross-modal abilities, but because of problems in training, for example, an extraordinary reluctance to reach into the aperture and then actively palpate an object. In a recent study at Oxford (Cowey and Weiskrantz, 1975), monkeys demonstrated this ability for cross-modal perceptual equivalence with a more appropriate procedure which required them to feel and eat sample objects in the dark and then select by vision alone the match to those samples.

We have also studied the chimpanzee's ability to respond on the basis of a "representation" of a visual sample by using life-size photographs of objects and degradations of the visual stimuli, e.g. half-size photographs, high contrast photographs (in which all textural and depth cues are eliminated), and even poor quality line drawings (Davenport and Rogers, 1971; Davenport, Rogers and Russell, 1975). Apes were successful with these.

Additionally, chimpanzees were successful in matching-to-sample when a 20-second delay was imposed between the sampling response and the availability of choices (Davenport, Rogers and Russell, 1975). Surprisingly, within the limits of our experiment, accuracy was no different with the delay than when the sampling and choice responses were simultaneous.

We have made a logical case for the transfer procedure being equivalent to matching-to-sample insofar as cross-modal perception is the object of study; however, the robustness of a phenomenon is increased if it can be demonstrated in different circumstances with different methods. Thus, the following experiment was done. Four chimpanzees which had been subjects in other reported cross-modal experiments were tested with a transfer procedure. They received repeated trials on haptic discrimination problems, each of which required the learning of the positive object. This was immediately followed by a single visual choice presentation of the two previously presented haptic discriminanda. After approximately 300 such problems in which the discriminanda had been repeated, they received 40 trials with completely new objects. Performance accuracy was at approximately the same level as in the delayed matching-to-sample experiment, which provides further evidence of the transsituational nature of the phenomenon in nonhuman primates.

Although, as previously mentioned, auditory-visual cross-modal perception of intensity or pulse trains in non-primates has been reported, we know of no studies of haptic-visual equivalence. To make sensible comparisons of taxa, methods, modalities and stimulus properties, clearly much more research is needed. Perhaps a neurological explanation for cross-modal perception may be found in some mechanism involving multi-modal cells, especially with simple, discrete, repetitive visual and auditory stimuli; however, the relevance of this for complex pattern recognition is not entirely clear. Indeed, there may be several mechanisms processing information of differing complexity and perceived by different peripheral sensory organs. Our approach so far has been to examine the behavioral phenomena involving complex stimuli without special attention to the neurological mechanisms.

In our view, cross-modal perception of equivalence requires the derivation of a modality-free "representation", "cognition" or concept of a stimulus or event. Animals that can have the same "representation", regardless of the means of peripheral reception, possess a great selective advantage in

coping with the demands of living, for example, in circumstances in which one modality may be operating sub-optimally
or when fast and accurate response is necessary on the basis
of limited information.

Le Gros Clark, among others, has given a prominent place
in hominid evolution to the increasing complexity of the brain,
which permits a functionally higher and more complete analysis
and synthesis of stimuli to which the end organs are already
receptive, and a lesser role to alterations in morphology of
the receptors per se (Clark, 1971). The methods of investigating cross-modal perception could provide powerful tools to
study the important and historically difficult issue of central
information processing in animals. The importance of the method for discovering and understanding phyletic similarities
and differences is clear.

REFERENCES

Blank, M. and Bridger, W. H.   Cross-modal transfer in nursery-
    school children. J. Comp. Physiol. Psychol. 58: 277-282,
    1964.
Bryant, P. E., Jones, P., Claxton, V. and Perkins, G. M.
    Recognition of shapes across modalities by infants.
    Nature, 240: 303-304, 1972.
Clark, W. Le Gros. The Antecedents of Man, 3rd ed. Quadrangle
    Books: Chicago, 1971.
Cowey, A. and Weiskrantz, L. Demonstration of cross-modal
    matching in rhesus monkeys, Macaca mulatta. Neuropsycho-
    logia 13: 117-120, 1975.
Davenport, R. K. and Rogers, C. M. Intermodal equivalence of
    stimuli in apes. Science 168: 279-280, 1970.
Davenport, R. K. and Rogers, C. M. Perception of photographs
    by apes. Behaviour 39! 2-4, 1971.
Davenport, R. K., Rogers, C. M. and Russell, I. S. Cross-modal
    perception in apes: Altered visual cues and delay.
    Neuropsychologia 13: 229-235, 1975.
Ettlinger, G. Analysis of cross-modal effects and their re-
    lationship to language. In: Brain Mechanisms Underlying
    Speech and Language, C. G. Millikan and F. L. Dailey,
    eds. Grune and Stratton: New York, 1967.
Ettlinger, G. and Blakemore, C. B. Cross-modal matching in
    the monkey. Neuropsychologia 5: 147-154, 1967.
Ettlinger, G. and Blakemore, C. B. Cross-modal transfer set
    in the monkey. Neuropsychologia 7: 41-47, 1969.
Geschwind, N. Disconnection syndromes in animals and man.
    Brain 88: 237-294 & 585-644, 1965.

Lancaster, J. B.   Primate communication systems and the emer-
     gence of human language.   In: Primates: Studies in
     Adaptation and Variability, P. C. Jay, ed.   Holt, Rine-
     hart and Winston: New York, 1968.
Over, R. and Mackintosh, N. J.   Cross-modal transfer of inten-
     sity discrimination by rats.   Nature (Lond.) 224: 918-
     919, 1969.
von Senden, M.   Raum-und Gestaltauffassung bei operierten
     Blindgeborenen vor und nach der Operation, J. A. Barth:
     Leipzig, 1932.
Ward, J. P., Yehle, A. L. and Doerflein, R. S.   Cross-modal
     transfer of a specific discrimination in the bushbaby
     (Galago senegalensis).   J. Comp. Physiol. Psychol. 73:
     74-77, 1970.
Wegener, J. G.   Cross-modal transfer in monkeys.   J. Comp.
     Physiol. Psychol. 59: 450-452, 1965.
Weiskrantz, L. and Cowey, A.   Cross-modal matching in the
     rhesus monkey using a single pair of stimuli.   Neuro-
     psychologia 13: 257-261, 1975.
Wilson, W. A. and Shaffer, O. C.   Intermodality transfer of
     specific discrimination in the monkey.   Nature (Lond.)
     197: 107, 1963.
Yehle, A. L. and Ward, J. P.   Cross-modal transfer of a speci-
     fic discrimination in the rabbit.   Psychon. Sci. 16:
     269-270, 1969.

* COGNITIVE PROCESSING IN GREAT APES (<u>PONGIDAE</u>)

Donald Robbins

Emory University
Atlanta, Georgia

In the classic work, <u>The Great Apes</u>, Robert M. Yerkes
(Yerkes & Yerkes, 1929) pointed out the relatively long history
of investigations into the capacities of the Great Apes
(<u>Pongidae</u>). Although much of the history appeared anecdotal
it did reveal that a large number of investigators were ponde-
ring the capacities of the orang-utan (<u>Pongo</u> <u>pygmaeus</u>), the
chimpanzee (<u>Pan</u> <u>troglodytes</u>), and the gorilla (<u>Gorilla</u> <u>gorilla</u>).
The evolutionary relationship between man and the apes is in
itself a compelling reason for studying the great apes since
the great apes approximate man, in a gross sense, more closely
than do the lesser apes, monkeys, and so on. Another reason
for studying great apes is that the processes operative in
them may be similar to those operating in young children particularly when the children are of a non-linguistic age. For
example, Davenport, Rogers and Rumbaugh (1973 found deficits
in discrimination performance in chimpanzees 12 years after
they had been reared in impoverished environments. Comparable
findings with children may make the apes a reasonable animal
model for attempting to reverse cognitive deficits resulting
from early experience.

At this centennial conference an entire session has been
devoted to investigations of language in the chimpanzee. In
its most general sense, language may be viewed as a manipulation
of symbolic material or information. An implicit assumption
often made is that the ability to manipulate symbolic information may be taken as evidence of cognitive processing. In this

* This research was performed at the Yerkes Regional Pri-
mate Center, Atlanta, Georgia, which is supported by Grant
RR-00165, from the National Institute of Health. I wish to
acknowledge the technical assistance of M. Blum, C. Bush,
C. Cochran, P. Compton, C. Gluck, S. Howard, D. Marcus, and
C. Scott.

manner, the use of language may be viewed as being a member
of the class of events that constitute evidence of cognition.
With regard to language, an assumption often implicit in most
theories of human memory and cognition, is that semantic or
linguistic information processing plays a critical role in
much of the phenomena observed in humans (e.g. Anderson and
Bower, 1973; Melton and Martin, 1972; Tulving and Donaldson,
1972). To be sure, the role of semantic or linguistic variables cannot be minimized in human beings. However, we wondered whether these variables were given too much weight, i.e.,
can phenomena, found with humans and assumed to be semantic-
ally based, be observed in non-human primates, in particular
Great Apes? To the extent that they do, it may indicate a
level of processing not requiring a semantic or linguistic
base. Thus the term cognitive processing may be used to refer
generally to the ability to manipulate symbols, whether they
are linguistic, imaginal or other forms. In this regard,
Davenport and Rogers (1970) and Davenport, Rogers and Russell
(1973) found that orang-utans and chimpanzees can discriminate between two objects based on either tactile or visual cues
and successfully match a sample presented in the other modality. This demonstration of cross modal transfer seems to indicate a phenomenon in apes which is based on a non-linguistic
process.

Medin and Davis (1973) recently reviewed the memory
literature on non-human primates and found it surprisingly
lacking. There has however, been a recent surge of interest
in animal memory (e.g., see Medin, Roberts & Davis, 1976), and
with regard to non-human primates, investigators have been
primarily studying short term memory in stumptail macaque
monkeys (Macaca speciosa) (Jarrad & Moise, 1970; Moise, 1970,
1976), and capuchin monkeys (Cebus apella), (D'Amato, 1973).
Little data exist, however, with regard to memorial and cognitive processing and capacities of the Great Apes.

Much of the memory research, until recently, has emphasized the limits of the apes' ability. Thus, investigators
often ask, using a delayed response task, what is the largest
delay that still yields performance above chance? Yerkes
(1928) himself, in his detailed study of memory with the
gorilla Congo, found increases in performance in a delayed response task as the delayed interval increased from 0 sec. to
600 sec. However, since the order of the delay conditions was
correlated with increasing amounts of training we do not know
what contribution, if any, the experience had. As a result,
it is unclear if the relation between performance and delay
reflects an effect of delay or improvement resulting from
additional training.

Despite the detailed descriptions found in Yerkes (1916, 1928), as well as Kohler (1925), systematic research investigating processes rather than establishing limits has typically not been done with great apes. The research reported here represents the beginnings of such a project.

In our first experiment (Robbins & Bush, 1973) we investigated memory in orang-utans, chimpanzees, and gorillas. We used a continuous multiple-problem two-choice simultaneous discrimination procedure. Each problem was presented for three trials, and the interval between trials (the intertrial interval) was varied and was filled with trials on other problems. Thus, the task can be viewed as a concurrent discrimination task.

Two gorillas, two orang-utans, and two chimpanzees were used in the experiment. Both gorillas were 9 years old, one was female (Oko) and the other male (Rann). The orang-utans were a 13-year-old male (Lembak) and a 14-year-old female (Tupa). Both of the chimpanzees were female; one was 8-years-old (Beleka) and the other 25-years-old (Debi). All but one were caged with another animal. All of the subjects had an extensive history of experimental experience and one (Debi) had been a subject in a sensory deprivation study as an infant.

A large-animal modified semiautomated Wisconsin General Test Apparatus was used (Rumbaugh, Bell & Gill, 1972). Basically, the apparatus consisted of two clear Plexiglas panels, arranged in a horizontal array, separated by a metal bar. Across the front of each panel were photoelectric beams, and a response was defined as the first side on which a photoelectric beam was broken. Behind each panel was a metal holder for the stimuli and above the panels was a receptacle for candy. The apparatus was mounted on a false door which replaced the door on the subjects' cage after the cagemate was removed. The stimuli were 10.16 x 13.97 cm. cutouts of magazine pictures chosen so that the stimuli were varied but relatively easily discriminable patterns (to the human observer).

Each subject was presented with a large number of two-choice simultaneous discrimination problems with three trials given on all of the problems. Each trial consisted of a presentation of two pictures, one behind each panel, thus defining a two-choice simultaneous discrimination task. The number of intervening trials (filled with presentations of other problems) between the first and second trial (0, 2, or 10) and the second and third trial (2 or 10) was varied factorially. Each subject was exposed to all six conditions. The conditions were presented to each subject in blocks approximately equating the number of times each condition appeared in the beginning, middle, or end of a block. A given condition was sometimes repeated within a block whenever the trial spacings permitted. As a result, subjects received different numbers of trials per condition. As a result, subjects received different numbers of trials per condition. For each of the problems one of the stimuli was arbitrarily called correct, and the number of correct stimuli on the right and left side was counterbalanced across conditions, although within a problem the positions were never changed. The blocks were presented to each subject in a different random order. All subjects were given preliminary training with three-dimensional objects on a two-object simultaneous task and achieved 100% correct performance within 40 trials. Then each subject was given 2-4 blocks on the first day of experimental training and eventually received 6 blocks per day for a total of 235 experimental problems. Except for the first day, each day began with a retention test on the problems from the previous days' first and last block. Since there were not many trials on the 24-hour retention tests, these data are not reported here. A noncorrection procedure was used and Hershettes (M & M type candies) were used as a reward.

Occasionally some subjects lapsed into a position habit, perhaps because of the difficulty of the task. When this occurred the experimental procedure was halted and the reward remained on the same side until the position habit was broken; then the experimental training session was continued. However, to avoid any possible bias in the data as a result of positional responding, all of the data collected on that day were eliminated from the analyses (range of 1-3 days).

The sequence of events for a typical trial was as follows: the experimenter turned the light on above the two stimulus pictures (by the time the experimental training was introduced all subjects responded within 1-2 sec.); the response terminated the light above the stimuli and reward was delivered if a correct response was made; the experimenter changed stimuli and set up reward conditions for the next trial during a 15-20 sec. intertrial interval.

The initial analyses revealed that there was little change in the proportion correct either over days or within days over block ($\underline{F} < 1$); this may have been because all of the subjects had had extensive experimental experience. As a result, all of the remaining analyses collapse the data over days and blocks within days. Because of the elimination of an entire day's data from the analysis if subjects' response pattern suggested a position habit, each condition for each subject may have had a different number of problems. The proportions of correct responses collapsed over all conditions were .506, 551, and .593 for Trials 1, 2, and 3, respectively (based on 200-230 total problems per subject).

An analysis of variance performed on the proportion of correct responses on Trial 1 revealed no initial differences between the conditions (used here as a "dummy" variable since the manipulations had not yet been introduced).

Since the individual subject data from this study has been reported previously (Robbins & Bush, 1973), and since the emphasis here is on a reanalysis of these data, we will only present the data averaged over the six subjects. In Figure 1 panel A, is shown the average proportion of correct responses on Trial 2 as a function of the lag (0, 2, or 10 intervening trials) between Trial 1 and Trial 2 (based on 63-87 problems

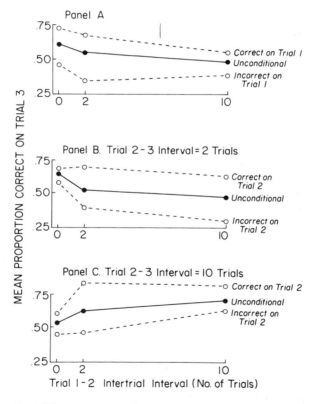

Fig. 1. Mean proportion of correct responses as a function of Trial 1-Trial 2 interval on Trial 2 (panel A); on Trial 3 when 2 trials intervened between Trial 2 and 3 (panel B) and when 10 trials intervened (panel C). The filled circles with solid lines represent the unconditional response proportions while the open circles with dashed lines represent the response proportions conditionalized on a correct or incorrect response on the immediately preceding trial.

per subject condition. In general, the proportion correct decreased as the lag increased, $F_{(2,6)} = 7.29$, $p < .05$. Most subjects were responding to close to a chance level at Lag 10 (range: .43-.55 proportion correct). In addition, when performance is conditionalized upon whether a correct response was made on Trial 1 or not, the function is steeper in the former case. These functions are not unlike short-term retention functions found by Medin (1969) for rhesus monkeys, by Moise (1970) for stumptail macaques, and those generally found for human subjects (e.g., see Bjork, 1969; Peterson, 1963).

In Figure 1 panels B and C are shown, the mean proportion of correct responses on Trial 3 as a function of the Trial 1-Trial 2 lag (based on 27-46 problems per subject per condition). Separate curves are shown, one in panel B for the massed (Lag 2) and one in panel C for the spaced (Lag 10) Trial 2-Trial 3 condition. For the spaced condition the data for almost all of the subjects revealed that the proportion of correct responses on Trial 3 increased as the Trial 1-Trial 2 lag increased. In addition, for all subjects, the proportion of correct responses on Trial 3 was greater for the Trial 1-Trial 2 lag of 10 intervening trials than for the lag of 0 intervening trials. This effect is often referred to as the "spacing" or "lag" effect in the human memory literature (see Bjork, 1969, Hintzman, 1974). It can also be seen that the mean proportion of correct responses decreased as the Trial 1-Trial 2 spacing increased when Trials 2 and 3 were massed (lag 2). This effect, which we will refer to as the "massing" effect, has also been found with human subjects; thus we find an interaction between the intervals between Trials 1 and 2 and Trials 2 and 3 as Peterson, Hillner and Saltzman (1962) reported for human subjects, as well as Robbins and Bray (1974). An analysis of variance performed on the proportion correct on Trial 3 revealed that the interaction was significant, $F_{(2,6)} = 27.79$, $p < .05$, although the Trial 1-Trial 2 main effect was not ($F < 1$), and the Trial 2-Trial 3 main effect fell short of statistical significance, $F_{(1, 3)} = 6.34$, $.05 < p < .10$. No obvious differences were found between the three species investigated.

Figure 1, panels B and C also shows performance on Trial 3 conditionalized upon Trial 2 performance. It can be seen in panel B for the short Trial 2-Trial 3 interval, that Trial 3 performance was relatively unaffected by the Trial 1-Trial 2 interval when a correct response had been made on Trial 2. In contrast Trial 3 performance, following errors on Trial 2, decreased as the Trial 1-2 interval increased. Panel C shows the data for the spaced (lag 10) Trial 2-Trial 3 interval. Here it can be seen that performance on Trial 3 increased as a function of the Trial 1-Trial 2 interval when a correct response had been made on Trial 2. When an incorrect response had been made on Trial 2, performance was essentially the same for the 0 and 2 intervening Trial 1-Trial 2 conditions, although a large increase was evident for the 10 intervening trial condition yielding a positively accelerated function. Although conditionalizing performance in this manner is not without the risk of contamination by item selection effects (e.g. stimulus preferences), the resulting fairly orderly functional relations seem to argue against these possibilities.

These data suggest to us that during the intertrial interval processing of the preceding information occurred. In this regard Wagner, Rudy, and Whittow (1973), using rabbits, found effects of an event occurring immediately after a Pavlovian conditioning trial. They interpreted their results in terms of interference with processing of information presented on the immediately preceding Pavlovian trial. Taken together these studies encourage the position that the intervening events affected the processing of prior events. In a subsequent study no specific events were presented during the interval between trials. This procedure which should minimize, if not eliminate, any interfering effects, should also maximize processing of the information presented on the immediately preceding trial. As a result, a factorial combination of Trial 1-Trial 2 interval (10, 20, 40, or 80 sec.) and Trial 2-Trial 3 interval (10 or 80 sec.) with unfilled intervals was used with three choice simultaneous visual discrimination problems. It should be noted that this method may be viewed as an adaptation of the classic Peterson and Peterson (1959) technique used to study human memory.

The two orang-utans and two chimpanzees used in the original study and the first experiment reported here were used again. A large animal wooden WGTA was also used. The stimuli were the two dimensional magazine pictures of the original study. Each subject was presented with a large number of three-choice simultaneous visual discrimination problems with three trials given on all of the problems. We switched to a three choice from a two choice task to avoid potential "floor" effects. Each trial consisted of a presentation of three pictures, one mounted on each slide panel, thus defining a three-choice simultaneous discrimination task. The amount of time (always an unfilled interval) between the first and second trial (10, 20, 40, or 80 sec.) and the second and third trial (10 or 80 sec.) of a problem was varied factorially. Each subject was exposed to all eight conditions. The conditions were presented to each subject in blocks approximately equating the number of times each condition appeared in the beginning, middle, or end of the block. For each of the problems one of the stimuli was arbitrarily called correct and

the position for each stimulus on each of the tree trials of
a problem were changed. The blocks were presented to each
subject in a different random order and the specific stimuli
assigned to a problem were randomly determined for each subject. All subjects were given preliminary training with
three-dimensional objects on a three object simultaneous task
and achieved 100% correct performance within 20 trials. Then
each subject was given one experimental block on the first day
of experimental training and received 1-2 blocks per day for a
total of 440 experimental problems (55 problems for each of
eight conditions). A noncorrection procedure was used and
Hershettes (M & M type candies) were used as a reward.

The sequence of events for a typical trial was as follows:
The experimenter positioned the apparatus and raised the front
cover, thus sliding the entire form board to within the subject's reach. The subject responded by pushing one of the
slides and thus exposing the foodwell. If a correct response
was made reward was present. After the subject removed the
reward the form board was pulled out of the subject's reach.
If an incorrect response was made no reward was present and
the form board was pulled out of the subject's reach immediately. This procedure was repeated for two more trials for a
given problem for the appropriate spacing conditions. After
three trials had been given on a problem another problem
followed. A one minute unfilled interval separated problems.

The initial analyses revealed that there was little
change in the proportion correct either over days or within
days over blocks, $F < 1$; again this was probably because all
of the subjects have had extensive experimental experience.
As a result all of the remaining analyses collapse the data
over days and blocks within days. The proportion of correct
responses collapsed over all conditions was .335, .339, and
.393 for Trials 1, 2 and 3, respectively (based on 440 total
problems per subject).

An analysis of variance performed on the proportion of
correct responses on Trial 1 revealed no initial differences
between the conditions, range of .30-.35, (used here as a
"dummy" variable since the manipulations had not yet been introduced), $F < 1$.

In Figure 2 is shown the proportion of correct responses
on Trial 2, for each subject as a function of the Trial 1-Trial
2 spacings (based on 110 problems per subject per condition).
The proportion correct appeared to be insensitive to the spacing and was fairly horizontal, revealing little if any learning, $F \ 1$. Most subjects were responding at close to a chance
level (.33) on Trial 2.

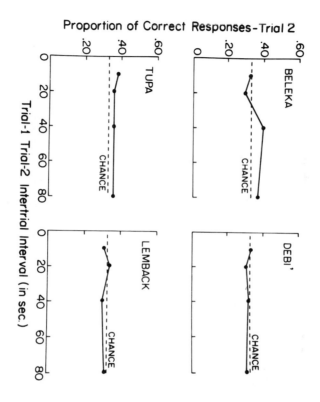

Fig. 2. Proportion of correct responses on Trial 2 as a function of the Trial 1-Trial 2 interval for each subject.

In Figure 3 is shown, for each subject, the proportion of correct responses on Trial 3 as a function of the Trial 1-Trial 2 interval (based on 55 problems per subject per condition). Two curves are shown for each subject, one for the massed (10 sec.) and one for the spaced (80 sec.) Trial 2-Trial 3 interval condition. For the spaced condition the data reveal that the proportion of correct responses on Trial 3 increased as the Trial 1-Trial 2 interval increased. In addition, for all subjects, the proportion of correct responses on Trial 3 was greater for the 80 sec. intervening Trial 1-Trial 2 condition than for the 10 sec. intervening condition. This

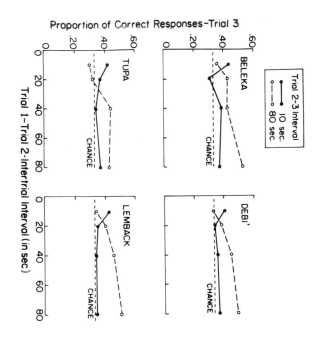

Fig. 3. Proportion of correct responses on Trial 3 as a function of the Trial 1-Trial 2 interval for the massed (10 sec.) and spaced (80 sec.) Trial 2-Trial 3 interval conditions for each subject.

effect is often referred to as the "spacing" or "lag" effect in the human memory literature (see Hintzman, 1974).

In Figure 3 is also shown the proportion correct on Trial 3 as a function of the Trial 1-Trial 2 interval when Trials 2 and 3 were massed (10 sec.). It can be seen that for all subjects the proportion of correct responses was fairly horizontal as a function of the Trial 1-Trial 2 spacing and once again appeared to indicate little or no learning. An analysis of variance performed on the proportion correct on Trial 3 revealed that the interaction was significant, F (3, 9) = 21.09, $p < .05$, $MS_e = .842 \times 10^{-3}$, as well as the Trial 1-Trial 2 main effect, F (3, 9) = 12.39, $p < .05$, $MS_e = 1.38 \times 10^{-3}$, although the Trial 2-Trial 3 main effect fell short of statistically significant, F (1,3) = 7.57, $.05 < p < .10$, $MS_e = .169 \times 10^{-2}$.

In Figure 4 is shown the mean performance (averaged over the four subjects) for Trial 3 for the 10 and 80 sec. Trial 2-Trial 3 intervals, in panel A and B, respectively. Performance conditionalized on correct and incorrect Trial 2 performance reveals a U-shaped function for the former and a decreasing function of Trial 1-Trial 2 interval for the latter for the 10 sec. Trial 2-Trial 3 interval, as shown in panel A. In contrast, as shown in panel B, when the Trial 2-Trial 3 interval was 80 sec., performance conditionalized upon correct Trial 2 performance was an increasing function of the Trial 2-Trial 3 interval and fairly horizontal for incorrect Trial 2 performance.

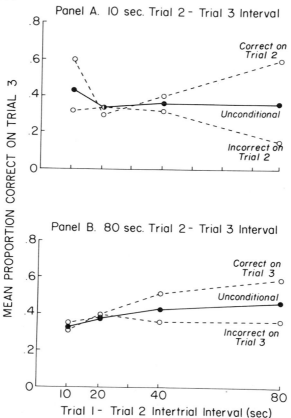

Fig. 4. Mean proportion of correct responses on Trial 3 as a function of the Trial 1-Trial 2 interval for the 10 sec. (Panel A) and 80 sec. (Panel B) Trial 2-Trial 3 intervals. The filled circles with solid lines represent the unconditional response proportions while the open proportions conditionalized on a correct or incorrect Trial 2 responses.

The results of this experiment are consistent with those of Robbins and Bush (1973) in finding a result similar to the lag effect. In a subsequent experiment another technique was used to investigate events affecting information processing. The subjects were given two trials on a large number of three-choice simultaneous visual discriminations. Some of the problems involved three dimensional objects while others involved two dimensional stimuli. Since it is well established, for non-human primates at least, that three dimensional object problems are easier to solve than two dimensional ones in learning set formation (Harlow & Warren, 1952) as well as in consective or concurrent simultaneous discriminations (Lynn, 1965), these problems were defined as "easy" and "hard" respectively. Further, if "easier" problems are interpreted as requiring less processing time to store information in contrast to a "harder" problem, then, if an easy problem intervenes between the two trials of a hard problem, performance on Trial 2 of the hard problem should be enhanced relative to a condition where the intervening problem is a different hard problem.

The orang-utans and two chimpanzees used in the initial study were the subjects in this investigation. Again a large animal wooden WGTA was used. Basically the apparatus consisted of a wooden form board, with three foodwells. Each foodwell was beneath a flat Plexi-glas slide which measured 5.08 x 20.32 cm. which had a Plexi-glas box and picture holder mounted on top of it. The slides were 5.08 cm. cutouts of magazine pictures chosen so that the stimuli were varied but relatively easily discriminable patterns (to the human observer). The three dimensional objects all fit in the Plexi-glas box and varied in color, size, and shape. They consisted of children's toys purchased at local thrift shops, electrical and electronic parts, small tools, and various items typically found in wood and electronic shops.

Each subject was presented with a large number of three-choice simultaneous discrimination problems with two trials given on all of the problems. Each trial consisted of a presentation of three objects, thus defining a 3-choice simultaneous visual discrimination task. A factorial combination of the type of problem, two or three-dimensional herein referred to as "hard" or "easy" problems, respectively, was varied and the type of problem intervening between the two trials of each problem was varied, so that four conditions resulted: two trials of an "easy" problem separated by a "hard" or "easy" problem and a "hard" problem separated by a "hard" or "easy" problem. Each subject was exposed to all four conditions. The problems were presented to each subject in a continuous sequence, approximately equating the number of times each condition appeared in the beginning, middle, or end of a daily session. Occasionally, filler problems were inserted to obtain the necessary sequences. For each of the problems one of the stimuli was arbitrarily called correct, and the number of correct stimuli in each of the three positions was equated across conditions. In addition, the positions within a problem were changed from Trial 1 to Trial 2. Each subject was given a different random order of problems and different stimulus combinations were randomly assigned to problems for each subject. All subjects were given pretraining on a three-dimensional and a two-dimensional task and achieved 100% correct performance within 40 trials on both tasks. Then each subject was given the experimental task. On the first day of experimental training each subject was given 20 problems and were then given 40 problems each day until a total of 280 experimental problems (70 problems for each of 4 conditions) were administered. A non-correction procedure was used and Hershettes (M & M type candies) were used as a reward.

The sequence of events for a typical trial was as follows: The experimenter positioned the apparatus and raised the front cover, activating a timer, and also sliding the entire form board to within the subject's reach. The subject responded by pushing one of the slides and thus exposing the foodwell and stopping the timer. If a correct response was made, a reward was found in the foodwell. After the subject removed the reward, the form board was pulled out of the subject's reach. If an incorrect response was made, the foodwell was empty and the form board was pulled out of the subject's reach immediately. A 30 sec. unfilled period followed, and then the next trial began. In this manner both choice behavior and choice response times were recorded.

The initial analyses revealed that there was little change in the proportion correct either over days or within days over blocks, $F < 1$; this may have been because all of the subjects have had extensive experimental experience. As a result, all of the remaining analyses collapse the data over days and blocks within days.

An analysis of variance performed on the proportion of correct responses on Trial 1 (range .31-.34) revealed no initial differences between the conditions, $F < 1$.

In Figure 5 is shown the proportion of correct responses on Trial 2 for each subject, as a function of the problem type (easy or hard) and the type of intervening trial (easy or hard). These proportions are based on 70 observations per subject per condition. It is clear from Figure 3 that for all subjects

three-dimensional problems were in fact easier than the two-
dimensional problems, so that the designation of easy and hard,
respectively, is supported by these data. The mean proportion
correct on Trial 2 averaged across all subjects and conditions
was .61 and .50 for the easy and hard problems, respectively.
An analysis of variance was performed on these data and
yielded a value just short of statistical significant for
Problem type, F (1, 3) = 10.00, MS_e = .967 x 10^{-3} , when the
required value is F (1, 3) = 10.13 for p = .05. However, the
effect of intervening trial task achieved statistical significance at the conventional .05 level, F (1, 3) = 34.78, $p <$
.05, MS_e = .692 x 10^{-3}. The interaction between the two variables also fell slightly short of statistical significance,
F (1, 3) = 8.40, .05 < p <.10, MS_e = .249 x 10^{-2}. With an
n = 4 the power of the test is obviously greatly reduced.
Further, since all four subjects show between a 20-50% improvement in performance on the hard problem when an easy problem
intervened in contrast to another intervening hard problem, we
feel that these effects can be taken as real. No other comparisons yielded significant effects, $Fs <$ 1.

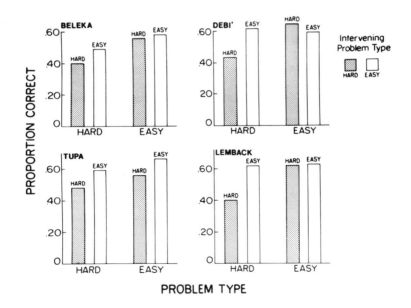

Fig. 5. Proportion correct responses as a function of
problem type and intervening problem type for each subject.

The response time data of Trial 2 was subjected to an analysis of variance comparing Problem type, Intervening trial problem type and Response (correct or incorrect) and revealed no significant effects, Fs < 1. The average response times on Trial 2 varied from 4-6 sec.

The results of this experiment reveal an enhancement in performance from Trial 1 to Trial 2 on hard problems when an easy problem intervened in contrast to when a hard problem intervened. This data may be taken to indicate that subjects were processing information during the intertrial interval and that the easy intervening task required less time to process, so that subjects may have begun processing the immediately preceding problem, which in this case was a hard problem, during the intertrial interval. In contrast, little if any processing time may have remained when the intervening problem was hard. Both the lag effect and the effect of an "easy" in contrast to a "hard" intervening task on performance on a hard problem support the general notion of the interfering effects of intervening material. However, some problems exist for this view. One puzzle is why subjects who had no specific task requirements during the intertrial intervals performed so poorly on Trial 2, and on Trial 3 when there was a short Trial 2-Trial 3 interval. These data indicate, when contrasted with Robbins and Bush (1973), that requiring attention to a specific intervening task appears to lead to similar subsequent performance than when no specific intervening task is required.

Nevertheless, these data may be viewed as additional evidence to support the view that Great Apes are capable of processing information when the information is no longer displayed in front of them. Although the theoretical explanation of these data may appear to be unclear the cognitive processing evidenced here is indicative of a cognitive capacity of the Great Apes. We will now turn to an attempt at a theoretical explanation.

Rather than consider a variety of theoretical explanations we will present one particular approach. We are doing this not because we are convinced that it is the most appropriate one, but rather it is presented as an example of a cognitive approach to information processing in the Great Apes.

The results of the first two studies suggests that relatively little learning occurs on the first trial, and when it does it appears to occur primarily when a response is followed by a reward. Thus, let us assume that the ape processes information slowly at first. However, the role of reward may be in part to lead to yet further processing. Further, we may assume that the longer the information is no longer in front of the organism (up to a limit of course) the more likely it is that additional processing has occurred and

Yerkes, R. M. Comparative Psychology Monographs, 1928, 5,
    whole no. 24.
Yerkes, R., & Yerkes, A. The Great Apes: A study of anthropoid
    life. New Haven: Yale University Press, 1929.

NATURALISTIC POSITIONAL BEHAVIOR OF APES AND MODELS OF HOMINID EVOLUTION, 1929-1976

Russell H. Tuttle

The University of Chicago
Department of Anthropology and
Committee on Evolutionary Biology
Chicago, Illinois

INTRODUCTION

Robert M. Yerkes' chief written contributions on the postures and locomotion of apes occur in the still informative reference volume titled The Great Apes, a Study of Anthropoid Life which he and his scholarly wife, Ada, published in 1929. In addition, Robert Yerkes inspired and otherwise directly supported the pioneer field studies of anthropoid primates by Henry W. Nissen (1931), Harold C. Bingham (1932), and Clarence Ray Carpenter (1934, 1935, 1938, 1940). Nissen's monograph on chimpanzees and especially Carpenter's monograph on lar gibbons contain much valuable information about naturalistic locomotor and other positional behavior of two apes which conspicuously influenced models of hominid phylogeny from the outset of the twentieth century (Tuttle, 1974).

In this tributary essay, I will sketch the state of knowledge about naturalistic hylobatid and pongid positional behavior from 1929 until 1976 and will comment briefly on its relevance to theoretical evolutionary anthropology. Gorillas are omitted because currently there is no evolutionary anthropological model premised on them. Bonobos are excluded because of the dearth of information about their naturalistic positional behavior.

POSITIONAL BEHAVIOR

The term positional behavior was proposed by Prost (1965) and resurrected by Rose (1973). It encompasses both locomotor behavior (movements from place to place) and postural behavior (when the relations between the subject and its external environment are relatively stable).

Yerkes and Yerkes (1929) focused primarily on the locomotor behavior of apes and, following Pocock (1905), devoted
special attention to the relative facility with which they
walk quadrupedally and bipedally on the ground and climb in
trees. They summarized information about captives as well as
free-ranging subjects and were generally careful to distinguish between them. They discussed available information
about preferred sitting, resting and sleeping postures but
presented virtually no information about foraging behavior and
feeding postures in natural settings. Their sections on
"feeding" are largely centered on what substances captives
will accept.

1. Hylobatid Apes

Yerkes and Yerkes (1929:53) opened their discussion on
the locomotor habits of hylobatid apes with the statement that
"Little is definitely known of the daily life of wild gibbons."
They inferred the following from anecdotal reports of natural
historians and travelers: (a) species of Hylobates moved
rapidly by arm-swinging from branch to branch instead of jumping or springing; (b) they were remarkably adept, even bird-
like, in the ability to change course during rapid arboreal
flight; (c) siamangs were less agile than the smaller gibbons;
(d) on the ground their bipedal ability and predisposition were
surpassed only by those of man among catarrhine primates; (e)
pongid-like quadrupedal knuckle-walking was undescribed for
gibbon or siamang; and (f) they slept upright in crotches of
trees assisted severally by hand and footholds.
Yerkes and Yerkes (1924:54) further remarked that systematic measurements of the distances covered by gibbons during
arm-swinging "are much to be desired." While one must with
some embarrassment report that these measurements still have
not been obtained, there has been notable expansion of knowledge about other aspects of hylobatid positional behavior
since 1929. The first major advance occurred in 1940 with
publication of Carpenter's meticulous descriptions of Hylobates lar in Thailand.
Carpenter confirmed that the basic stances of gibbons
prior to locomotion are orthograde ones. Thus, locomotion may
be launched from suspensory postures below branches or bipedal
postures atop them. He also confirmed that sitting postures
are common while sleeping, though gibbons also sleep supine or
on their sides on wide branches.
Carpenter (1940, pp. 66-69) provided the first detailed
description of brachiation based on analyses of slow motion
films. He also noted that brachiation may be mixed with bipedal running if large, solid branches, four to six inches in

diameter, occur on the pathway. He estimated that gibbon locomotion consists of approximately 90 percent brachiation and 10
percent bipedal walking (p. 73) and noted that "There are many
variations of patterns from pure brachiation on one extreme to
pure upright walking on the other extreme." (p.70). On slender
terminal branches, brachiation was used exclusively. Bipedalism most commonly occurred on solid horizontal supports.
Carpenter stated that gibbon locomotion is phasic as to speed
and that bursts of speed are sporadic and of short duration
(1940, p. 78). Normally they moved through the forest at a
leisurely pace.

Carpenter (1940, p. 74-75, 70) emphasized the adeptness
with which gibbons climb vertical surfaces like tree trunks
and vines. He noted that elongate forelimbs and the deep
clefts between the thumb and the other manual digits greatly
facilitated their vertical climbing.

Like many other field naturalists, Carpenter loosely employed the terms leaping and jumping when referring both to
forelimb propelled and hindlimb propelled translocations. He
did not clearly distinguish between them. One senses that his
statement "A normal part of gibbon locomotion is that of jumping," chiefly refers to forelimb propelled translocations. But
some passages in his field notes indicate that hindlimb propulsion also occurred; for example:

"With extreme rapidity she swung underhand then on top
of the limb and then, with only a slight loss of time
and momentum, jumped outward and downward 30 feet to an
adjacent tree top." (1940, p. 76).

Carpenter (1940, p. 81-85) presented the first description of naturalistic food collecting postures in hylobatid
apes. He concluded that brachiation is an adaptation useful
in the frugivorous habits of gibbons, enabling them to reach
foods at the ends of springy branches overhead. They frequently gathered fruit in hands or feet and carried them to
places where they could be eaten more conveniently (p. 82).
Occasionally they ventured briefly to the ground for fallen
fruits.

A quarter of a century after Carpenter's classic study,
systematic field studies of hylobatid apes were resumed as
part of the general renaissance of primate studies which extends to the present day. In West Malaysia, Ellefson (1968,
1974) conducted a much longer study than Carpenter's on lar
gibbons. He basically confirmed Carpenter's observations on
their positional behavior. But Ellefson also augmented previous observations and speculated more freely than Carpenter
did about possible evolutionary meanings of his data.

Ellefson (1974, p. 71) concluded that the locomotor,
feeding and resting postures of his subjects did not conform

to strict patterns. For instance, he stated that "Brachiation, meaning solely arm-swinging, should not be invoked to characterize gibbon locomotion." (1974, p. 66) because most movements over more than a few yards include variable amounts of bipedal running, dropping to lower strata, vertical climbing, bridging, hoisting, and hauling, in addition to arm-swinging. He confirmed that most locomotion is leisurely and is in the context of foraging instead of major translocations (1974, p. 66). Although siamangs were not his focal subjects, encounters with them led Ellefson (1974, p. 25) to the impression that they move farther and faster than lar gibbons when being tracked. His lar gibbons "used all manner of different routes" instead of well-worn pathways (1974, p. 69). Bridging actions, whereby the subject pulled branches toward itself with the feet and then slowly transferred onto the new support, were common (1974, p. 68).

Ellefson (1974, p. 66) stressed that lar gibbons are excellent jumpers and climbers. But he did not designate whether the forelimbs or the hindlimbs were the chief propellant organs during "jumps" and "leaps". He remarked that they "land on their hands" at the end of leaps over twenty feet (1974, p. 68).

Ellefson (1974, p. 56) noted that feeding was most commonly conducted from suspensory postures in the terminal branches. The subject might hang by one hand and bring foods to mouth with the other hand. The feet might hold or pull the branch being fed upon. Feeding postures varied considerably depending upon the tree species and plant part eaten. The supporting forelimb changed often in order to reach new food sources and probably not because of fatigue. Ellefson's subjects also commonly fed while sitting on branches. Ventures to the ground were rare and usually were executed for choice morsels like stick insects.

When resting, lar gibbons usually sat hunched over and grasped neighboring branches with their hands. They also lay supine or on their sides for long periods while resting and grooming (1974, p. 71). Ellefson (1974, p. 73-74) concluded that his subjects slept sitting on large horizontal branches though a variety of supine and other nonprone reclining postures preceded sitting postures prior to sleeping. When he returned to sleeping sites early in the morning they were in the same or similar positions.

Acrobatic arm-swinging and other suspensory behavior are conspicuous components of the intraspecific territorial conflict displays of male lar gibbons (Ellefson, 1968, 1974). The great amount of display behavior observed by Ellefson may be due partly to the superabundance of males in the isolated population which he studied. But display swinging is a well documented feature of hylobatid adaptive complexes.

Ellefson concluded that the anatomical specializations of hylobatid apes which commonly have been associated with brachiating locomotion are probably better viewed primarily as adaptations for foraging and feeding in terminal branches.

At a different locality in West Malaysia, Chivers (1972, 1974) conducted the first systematic long-term study of siamangs. He compared their behavior with his and Ellefson's data on lar gibbons. Chivers confirmed the observations of most other observers (McClure, 1964; MacKinnon, 1974b; Miller 1942; Papaiouannou, 1973; Tenaza and Hamilton, 1971; Wallace, 1869) that siamangs generally move more slowly and cautiously, and less acrobatically than smaller gibbons do. The members of siamang family groups commonly proceeded together single file along well-known pathways (Chivers, 1972, p. 114). Chivers (1972, p. 129) summarized the locomotion of gibbons as "fast, light-jump" and that of siamangs as "slow, heavy-bridge" with little further explanation except to state that "Gibbons flit rapidly through the trees, and perch on small branches in a hunched position; siamang are slower and more deliberate in their movements (although capable of the spectacular locomotor feats frequently performed by gibbons), and rest draped across large branches or in forks where they can prop themselves." (1972, p. 116). He also confirmed (1972, p. 117, 1974, p. 238) that, unlike lar gibbons, female and other members of siamang groups join the adult males in display-swinging.

In Chivers' (1974) monographic "quantitative description" of siamang socioecology, locomotion and posture were slighted by comparison with information on ranging and social behavior. He remarked that the movements of siamangs through the trees are very similar to those of chimpanzees and orangutans (p. 295). But he did not present quantitative data on the frequencies of bipedal versus suspensory locomotion, leaping, vertical climbing, and bridging, or of sitting versus reclining rest postures.

It may be inferred from Chivers' (1974, p. 23, 49) discussions that his subjects slept sitting in the terminal branches of tall emergent trees; the dimensions and flexibility of the branches were not stated. Chivers (1974, p. 116) noted that, like lar gibbons, siamangs fed from suspensory or sitting postures according to the context in which foods were located. Thus, seasonal fruits which were densely clustered in terminal branches of relatively small trees were collected from suspensory postures; flowers and new leaves were more commonly collected from sitting postures. Because siamangs have a much greater predilection for folivory (Chivers, 1972, p. 120; 1974, p. 121, 127), they probably feed more often while sitting than lar gibbons do. Chivers (1974,

p. 116, 123) also noted that the largest animal in a group of
siamangs, a male, showed the greatest tendency to sit while
feeding.   Chivers (1974, p. 123) concurred with Ellefson that
"brachiation might be a feeding rather than a locomotor adaptation" and that the siamang probably represented the maximal
size capable of feeding while suspended in terminal branches.

The gap in knowledge about naturalistic positional behavior of siamangs was narrowed considerably by Fleagle (1976),
who studied them with Chivers in West Malaysia. He also observed sympatric lar gibbons during fewer hours using the
same quantitative approach which he employed with the siamangs.
In addition to confirming that eye-catching ricochetal arm-
swinging between branches is less frequent in adult siamangs
than in lar gibbons and that siamangs travel regularly on
specific routes, Fleagle (1976) noted the incidences and contexts of brachiation, climbing (including bridging), leaping,
and bipedalism in the two species.

Fleagle gained the impression that in siamang, climbing
was forelimb-dominated. He also stated that "the leaping
abilities of the siamang and hence their abilities to cross
discontinuities in the forest canopy are poor compared with
the abilities of other Malaysian higher primates." He observed that leaps were always from higher to lower levels and
usually between terminal branches no more than ten meters
apart horizontally. Commonly, the subject pumped in place,
then pulled forward with its forelimbs to launch itself. The
feet usually released the base branch last. But Fleagle
ascribed very little thrust to the hindlimbs during leaping.
Vertical drops were also common. Landing sites were grasped
with all hands and feet.

Fleagle (1976) found that in siamangs, brachiation accounted for 51 percent, climbing for 37 percent, bipedalism
for 6 percent, and leaping for 6 percent of bouts of travel
over relatively long distances. Computing the proportion of
distance covered by each of the modes, Fleagle found that
brachiation was employed over nearly 67 percent, climbing
about 25 percent and leaping and bipedalism together less than
10 percent of distance traveled. During travel, brachiation
was virtually confined to boughs and branches greater than
two centimeters in diameter. Climbing and leaping occurred
on a wide variety of substrates, including twigs less than
2 centimeters in diameter. And bipedalism was confined to
sturdy horizontal branches.

During travel, lar gibbons used more leaping (15 percent)
and less climbing (21 percent) than siamangs did and more of
their brachiation (56 percent) was ricochetal (23 percent of
total travel bouts). Fleagle (1976) inferred that bipedalism
is not significantly more common in siamangs than in smaller

gibbons (8 percent of travel bouts in <u>Hylobates</u> <u>lar</u>). Generally, lar gibbons engaged in the four different modes of locomotion on supports which were of similar sizes to those used
likewise by siamangs.

Siamangs employed climbing much more frequently (74 percent of bouts) than brachiation (23 percent) during foraging
translocations (termed "locomotion during feeding" by Fleagle,
1976). Bipedalism (3 percent of bouts) and leaping were rare
during foraging. Brachiation and bipedalism occurred on
sturdy supports while climbing commonly occurred on small
substrates where the foods often were located. Fleagle (1976)
concluded that climbing is "the siamang's most effective pattern of movement in a small branch setting."

Fleagle (1976) found that foraging lar gibbons use
brachiation more (45 percent of bouts) and climbing less (51
percent of bouts) than siamangs do. Bipedalism and leaping
each accounted for 2 percent of bouts of foraging movement in
lar gibbons studied by Fleagle.

Fleagle (1976) presented quantitative data on feeding
postures and nature of supports; he did not discuss sleeping
and resting postural behavior. He recorded 849 occurrences
of suspensory feeding and 527 episodes of sitting feeding in
siamangs. Forty-eight percent of suspensory feeding postures
occurred while siamangs were supported from relatively small
twigs; 49 percent occurred while they were hanging below
branches between 2 and 10 centimeters in diameter. Only 7 percent of sitting feeding episodes occurred among twigs, while
78 percent were performed from branches. The remaining 15
percent of sitting feeding was executed from supports greater
than 10 centimeters in diameter. Sixty-five percent of fruit
collection occurred from suspensory postures and 35 percent
from seated postures. Per contra, 69 percent of leaf collection occurred while sitting and 31 percent while hanging.

In 72 percent of siamang suspensory feeding episodes,
three appendages grasped supports. In 19 percent of suspensory
feeding episodes, two appendages were employed and in only 9
percent of episodes one appendage held onto supports. Siamangs,
also grasped supports while sitting feeding: they used three
appendages during 69 percent of episodes, two appendages during 28 percent of episodes, and one appendage during 2 percent
of episodes. Although much more limited data (48 episodes)
were collected on lar gibbons, Fleagle (1976) demonstrated
that they used three appendages for support during only 40
percent, two appendages during 27 percent, and one appendage
during 33 percent of suspensory feeding episodes. He inferred
that siamangs exploit the same small branch regions that
gibbons utilize by employing more hands and feet to support
themselves during feeding.

2. Orangutans

     Yerkes and Yerkes (1929, p. 112) commenced their dis-
cussion of locomotion in the "strictly arboreal" orangutan
by stating that it "differs conspicuously from the gibbon in
its slow, deliberate, cautious, almost slothful movements."
They relied primarily upon anecdotal accounts by Schlegel and
Muller (1839-1844), Wallace (1869), Hornaday (1879, 1885),
and other nineteenth century hunter-naturalists. They determined that orangutans rarely brachiate (i.e., travel by
arm-swinging unassisted by pedal grasps), spring or jump
(Yerkes and Yerkes 1929, pp. 117, 537, 556). Instead of brachiating, orangutans would climb with free use of hindlimbs
and forelimbs. Bridging was a common mode of transferring from
tree to tree. Captive orangutans were portrayed as relatively
slow quadrupeds on the ground. Yerkes and Yerkes (1929, p.
555) indicated that when on the ground, weight usually rested
on the outer edges of the feet and "on knuckles". They did
not distinguish the terrestrial hand postures of orangutans
from true knuckle-walking hand postures of the African apes
(Tuttle, 1967). Reviewing somewhat equivocal observations on
the bipedalism of captives, Yerkes and Yerkes (1929, p. 116)
concluded that the occasional bipedal stances and steps of
orangutans are "extremely difficult and fatiguing". They
cited no instances of arboreal bipedal progression. They
ranked orangutans as the least bipedal of all apes (p. 537).
     Yerkes and Yerkes (1929, p. 117) concluded that crouching
was a common daytime resting posture of orangutans and that,
unlike gibbons, weight rarely rested on their "ischial regions".
They reported that orangutans slept in nests lying supine or
on their sides (p. 122). Yerkes and Yerkes (p. 118) regarded
the elongate forelimbs of orangutans as greatly advantageous
for reaching food on branches which could not support their
full weight.
     Although Carpenter (1938), Schaller (1961), Harrisson
(1962) and other field behavioralists surveyed orangutans in
Sumatra and Borneo, and even though several respectably long
projects have now been conducted (Davenport, 1967; Galdikas-
Brindamoor, 1975; Horr, 1975; MacKinnon, 1971, 1973, 1974a,
1974b; Rijksen, 1975; Rijksen and Rijksen-Graatsma, 1975;
Rodman, 1973), there has been remarkably little advance in our
knowledge of positional behavior in Pongo since Yerkes' and
Yerkes' compendium. Only Schaller (1961), Davenport (1967)
and MacKinnon (1971, 1974a, 1974b) specifically discussed arboreal locomotor behavior in Pongo. Detailed descriptions,
like those of Carpenter (1940) on lar gibbons, are rare and
quantitative data, like that which Fleagle (1976) provided
about West Malaysian apes, is non-existent.

Recent observations on wild orangutans and naturalized captives confirm that their chief mode of locomotion is versatile climbing, including much cautious bridging, vertical ascents and descents, hoisting, hauling, and occasional pedal assisted arm-swinging and quadrupedal suspensory movement beneath branches (Schaller, 1961; Davenport, 1967; MacKinnon, 1971, 1974a, 1974b). When orangutans could not reach new supports across gaps in the canopy directly, they swayed the base supports to close the distance. MacKinnon (1974b, p. 30) remarked that definite arboreal "highways" were used by his wild subjects.

All observers who specifically discussed locomotion concur that unassisted brachiation is rare and occurs only over short distances. MacKinnon (1974a, p. 42) remarked that "Orangutan brachiation lacks the speed and flow of the specialized ricocheting brachiation of gibbons and siamangs and also differs in that the arms are swung overhead rather than underarm."

Like brachiation, jumping and vertical dropping over notable distances were confirmed to be quite uncommon among orangutans. MacKinnon (1971, p. 160; 1974a, p. 42) commented that leaping was only exhibited by fleeing subjects. He also witnessed "tumble descents" in which the fleeing subject fell rapidly through the foliage while briefly grasping and releasing supports with hands and feet (MacKinnon, 1971, p. 160; 1974a, p. 42). Schaller (1961) observed one jump over three feet to a lower branch. Davenport (1967, p. 252) observed no leaping or jumping. However, two of his subjects executed dramatic "dives" in which they held onto a branch with their feet and lunged or fell forward so that ultimately they hung by their feet alone.

On large horizontal and slightly inclined branches, orangutans normally walked quadrupedally (Schaller, 1961; MacKinnon, 1971, p. 157; 1974a, p. 40). Arboreal bipedal locomotion was not observed by Schaller or MacKinnon. Davenport (1967, p.251) mentioned that "under appropriate conditions" his subjects "appeared to be capable of progressing ... bipedally on top of a branch while holding vines and branches with one or both hands for support and balance ..." But he did not indicate how frequently this behavior actually occurred.

The only recent observations which add a really new dimension to our knowledge about locomotion in orangutans are those of terrestrial travel by adult males. Quadrupedal terrestrial locomotion by wild orangutans has been reported by deSilva (1971), MacKinnon (1971, 1974a, 1974b), Horr (1975), and Galdikas-Brindamoor (1975). In some instances, terrestrial locomotion was used to cross man-made discontinuities in the forest (e.g.; deSilva, 1971, pg. 74; Galdikas-Brindamoor, 1975,

p. 444). But in other instances, the context appears to be
wholly naturalistic. The terrestrial locomotion is generally
associated with large size (in some cases including obesity)
and reduced agility of adult males. According to MacKinnon
(1974a, pg. 43), "orangutans generally prefer to take an arboreal route even when a terrestrial route would seem easier."
Sumatran tigers and clouded leopards could be significant men-
aces to lone infirm adult males or smaller orangutans. Rijksen
and Rijksen-Graatsma (1975) lost seven naturalizing captives to
a clouded leopard, probably because of attacks on the ground.

     Schaller (1961), Harrisson (1962), Davenport (1967) and
MacKinnon (1971, 1974a, 1974b) provided the chief new information about feeding and resting postural behavior in Pongo.
Davenport (1967, p. 254) simply stated that orangutans fed in
almost any posture. Schaller (1961) did not always specify
which postures were used during feeding. But he remarked that
"Hanging by one arm and foot while reaching for food is not
unusual, and once a subadult hung two to three seconds by its
feet alone." (p. 78). Harrisson (1962 - p. 78) observed a
young orangutan hanging by one forelimb while holding a recently collected durian fruit with both feet and the other
hand. She also described an orangutan standing bipedally on
a branch and gripping overhead with one hand as it drew sprays
of leaves to mouth with the other hand. Concurrently, a second subject stood with thighs abducted on twisted branches and
held a vine as it alternately scraped tree bark with its teeth
and probed with a finger (Harrisson, 1962, p. 68). Harrisson
(1962, p. 71) and MacKinnon (1971, p. 165) observed that large
fruits and clusters of smaller food were sometimes carried to
large branches or nests where they were eaten. MacKinnon
(1974a, p. 43) provided the best general summary of feeding
postural behavior in Pongo, as follows:

     "Feeding animals usually sit on a firm branch holding
     on above with one arm whilst gathering and handling
     food with the other. At other times they hang out
     from, or beneath, a branch suspended by one arm and
     one leg. The free arm and leg are used for gathering
     and holding food. Sometimes animals hang by both
     hooked feet and use both hands to collect and open
     fruit."

MacKinnon (1974a, p. 117) also described an adult female
and youngster clinging vertically to the trunk of a tree as
they stripped and chewed its bark. He also observed an adult
male, estimated to weigh 200 pounds, hanging unimanaully below
a branch as it picked and ate ripe fruit (1974a, p. 121).

     Davenport (1967, p. 256) reported that his subjects
characteristically slept on their sides or supine. But again
it was MacKinnon (1974a, p. 43) who provided the best overview

of resting postures in <u>Pongo</u>:

> "Animals resting in trees, nests or on the ground
> often sat upright, sometimes holding an overhead
> branch with one arm. Some animals lie down on their
> backs or fronts with their limbs dangling over the
> branches. Orangutans can remain suspended between
> two upright supports with their legs taking much of
> their weight."

3. Chimpanzees

In the introduction to their section on locomotion of
chimpanzees, Yerkes and Yerkes (1929, p. 213) apologized that
"As usual, there are informational fragments from many sources,
but nowhere completeness of description." They inferred that
chimpanzees are "eminently arboreal" and cited their climbing
skills. While acknowledging that chimpanzees "can swing themselves rapidly from branch to branch and may on occasion as
it were throw themselves or fall through the air from one hand-
hold to another", Yerkes and Yerkes (1929, p. 214) remarked
that their mode of arboreal flight "falls far short" of that
of gibbons. They concluded that chimpanzees are equally well
adapted to terrestrial and arboreal contexts. They noted that
terrestrially chimpanzees normally progress quadrupedally on
knuckled hands and plantigrade feet and that although they are
"capable of standing and walking erect for short times and
distances," bipedalism is relatively difficult for them (1929,
p. 215). Their accounting of jumping by chimpanzees is somewhat ambiguous. In the text (p. 215) they stated that "Whereas jumping by leg propulsion when on a tree is an exceptional
occurrence...similar jumping on the ground is commonly observed." But in their synoptic table (p. 556) they indicated
that the chimpanzee "Jumps freely and skillfully in trees or
on the ground." They may have included forelimb propelled
translocations and vertical dropping among arboreal "jumps".

Yerkes and Yerkes (1929, p. 556) noted that chimpanzees
sleep lying supine or on their sides in arboreal nests. They
made no definitive statements about other postural behavior of
chimpanzees.

Like Carpenter (1940) on lar gibbons, Nissen (1931) obtained much useful information on the positional behavior of
chimpanzees during a relatively short field study. He reported that although Guinean chimpanzees spent a considerable
portion of their waking hours foraging, feeding and resting
in trees, and although they slept in tree nests at nights,
they generally moved from one site to another on terrestrial
pathways (pp. 34-35). Adults climbed up trees, even ones
measuring four or five feet in diameter, with ease. Descents

on tree trunks were notably more difficult to execute and required a longer time than ascents. Chimpanzees climbed down
large trees feet first. Some trees of small diameter were descended head first (Nissen, 1931, p. 34). Nissen (1931, p.
34) reported that a "much more common way" for a chimpanzee to
quit trees with branches fairly close to the ground was to walk
onto a low branch which bent under its weight and then either
step off or drop to the ground after hanging momentarily by
the forelimbs alone.

     Nissen's subjects did not move very far horizontally in
the canopy (p. 34). During some short translocations between
trees, they sometimes executed "jumps" which chiefly carried
them downward. Commonly, they hung and swung by their forelimbs and then let go of the base branch on a forward swing
(p. 35). Nissen also saw one female execute a "short jump"
from one tall tree to another after "bending rapidly at knees
about 6 times" (pp. 20-21). Nissen commented that while in
a fromager tree 150 feet above the ground a chimpanzee always
travelled quadrupedally on branches (p. 21). Although he saw
chimpanzees stand bipedally for brief periods on the ground,
he never saw them walking bipedally (p. 37).

     Nissen (1931, p. 67) described chimpanzee feeding as
follows: "sitting, squatting, reclining, or standing on a
limb ..., one hand grasping the same limb or a neighboring
one (usually one overhead)", as it reached out to the fruit
with its free hand. He did not mention chimpanzees feeding in
suspensory postures. Sometimes they used their feet to bend,
break or hold branches while feeding from them. Adult chimpanzees rested on branches or crotches of trees or, more often,
on the ground. In trees they reclined supine or on their
sides, generally with manual or pedal grasps of branches overhead (p. 31). As far as Nissen could determine, chimpanzees
slept supine or on their sides in night nests with their limbs
drawn up close to their bodies (p. 45).

     During the past fifteen years Nissen's observations have
been corroborated and, to some extent, augmented by several
long-term studies and surveys of chimpanzees in Western,
Central, and especially East Africa. However, in general,
recent observers of chimpanzees focused systematically on
positional behavior about as seldom as students of orangutans
did.

     All studies confirm that while chimpanzees preferentially
feed in trees and nest arboreally at night, they usually travel by knuckle-walking on the ground. This is true of chimpanzees in a wide variety of habitats ranging from dense forests to relatively open mosaics of woodland, savanna, and
riverine forest. (Goodall, 1962, 1963a, 1963b, 1965; Lawick-
Goodall, 1968; Kortlandt, 1962, 1968, 1972; Reynolds, 1965;

Reynolds and Reynolds, 1965; Izawa and Itani, 1966; Itani and
Suzuki, 1967; Bournonville, 1967; Rahm, 1967; Sugiyama, 1968,
1969, 1973; Suzuki, 1969; Izawa, 1970; Jones and Pi, 1971;
Albrecht and Dunnett, 1971; Kano, 1972). Lawick-Goodall,
(1968, p. 177) noted that chimpanzees "often help themselves
up steep slopes by hauling on tree trunks or low bushes" and
that as they descended steep slopes they "frequently" crutch-
walked, i.e., placed both hands on the ground and swung the
lower body forward between their abducted arms.

According to Lawick-Goodall (1968, p. 177) "Chimpanzees
frequently stand upright in order to look over long grass and
other vegetation" and they "frequently walk bipedally for short
distances" in tall grass, on wet ground or when carring food
and run bipedally during terrestrial branch waving displays.
They also exhibited a "bipedal swagger" during threats,
greeting and courtship displays (Lawick-Goodall, 1968, p. 177).
The frequencies of some of these bipedal events probably were
increased by the artificial feeding situation at Gombe.
Kortlandt (1962, 1967, 1968, 1972 Kortlandt and Kooij, 1963)
also was impressed with the variety and frequency of terrestrial bipedal behavior of wild chimpanzees. But again he
commonly observed them in the vicinity of plantations, pro-
visions or experimental objects. Reynolds (1965, p. 62).
regarded terrestrial bipedal walking as "rather uncommon" in
chimpanzees of the Budongo Forest. Bipedal standing "to get
a better view" was more common than bipedal movement on the
ground.

At Gombe chimpanzees "usually" leaped quadrupedally or
bipedally across streams or small gulleys (Lawick-Goodall,
1968, p. 177). Kortlandt (1962, p. 8) observed chimpanzees
jumping bipedally over six feet "from a standing start" to
cross a stream on a plantation. Reynolds and Reynolds (1965,
p. 439) also observed quadrupedal and occasionally bipedal
jumps across streams. Per contra Izawa and Itani (1966, p.148)
inferred that chimpanzees in the Kasakati Basin ordinarily
waded streams.

The adeptness with which chimpanzees climb vertical tree
trunks and boughs is attested by all observations. Indeed
Kortlandt (1968) emphasized that the long forelimbs and large
hands of chimpanzees enable them to obtain food from isolated
and emergent trees which do not have lianas extending to lower
strata and which thus could not be climbed by monkeys. Goodall
(1965, p. 437) commented that as subadults, chimpanzees "move
easily and quickly in trees but as they attain maturity arboreal locomotion becomes slow and careful unless the animals
are frightened or excited." Some branches are gripped with
hands and feet. But if the branches are wide, knuckle-walking

may be employed (Reynolds, 1965, p. 62; Lawick-Goodall, 1968,
177-178;   Albrecht and Dunnett, 1971, p. 18; Kortlandt,
1972, p. 15). Bridging and bending branches into juxtaposition
are reported to be common modes for chimpanzees transferring
from one tree to the next (Goodall, 1965, p. 439; Izawa and
Itani, 1966, p. 127). All observers noted that travel through
the canopy is limited to short distances.

Brachiation, "jumping" or "leaping", and bipedal locomotion sometimes occurred in trees. But it is impossible to
discern actual frequencies of these behaviors in Pan troglodytes from available reports. Goodall (1965, p. 440) stated that
"Brachiation for short distances is common" and mentioned it
as a prelude to bridging (Lawick-Goodall, 1968, p. 178).
Reynolds and Reynolds (1965, p. 384) listed brachiation along
branches as "common". But elsewhere Reynolds (1965, p. 59)
stated that he saw brachiation occasionally. Izawa and Itani
(1966, p. 127), Albrecht and Dunnett (1971, p. 20) and Jones
and Pi (1971, p. 61) simply state that they sometimes observed brachiation.

Most observers reported seeing short arboreal jumps.
They did not always mention whether they were chiefly propelled
by forelimbs or hindlimbs. Goodall (1965, p. 440) described
some jumps that were powered by the hindlimbs. Playing youngsters, displaying adults, and fleeing chimpanzees of all ages
also executed longer "leaps", at least some of which were probably vertical drops (Reynolds and Reynolds, 1965, p. 384;
Reynolds, 1965, pp. 58-61; Izawa and Itani, 1966, p. 127;
Jones and Pi, 1971, p. 61; Albrecht and Dunnett, 1971, p. 20).
Chimpanzees may grasp landing sites quadrupedally or with
their hands alone (Goodall, 1965, p. 440).

Reynolds and Reynolds (1965, p. 384) noted that occasionally chimpanzees walked short distances bipedally on branches
while grasping supports overhead with their hands. Reynolds
(1965, p. 62) also stated that "they would often walk bipedally
along "stout branches" in this manner. Rahm (1967, p. 203)
stated that chimpanzees fleeing from net hunters in Zaire very
often ran "upright like gibbons over branches" and jumped distances "they would not even try under normal conditions."
Other observers did not discuss bipedal locomotion.

Goodall (1963, p. 41) illustrated five arboreal feeding
postures of chimpanzees; the subjects were crouched or sitting
on branches or crotches, usually holding a higher branch with
one hand and feeding with a free hand. Further, she mentioned
that they stood on branches while feeding and sometimes hung
from one hand or a hand and foot in order to reach food at the
ends of branches (Lawick-Goodall, 1968, p. 185). Reynolds
(1965, p. 59) also observed unimanual suspensory feeding by
his subjects. Reynolds (1965) and other students of Budongo

chimpanzees (Sugiyama, 1968, 1969, 1973; Suzuki, 1971) have illustrated sitting feeding, especially associated with figs which grow on trunks and large boughs. Jones and Pi (1971, p. 61) remarked that in Rio Muni chimpanzees "sat and reclined frequently, as well as stood upright occasionally, on large limbs of trees." It is unclear whether they were feeding, resting or engaged in other behavior.

Lawick-Goodall (1968, p. 201) illustrated various sitting and reclined rest postures of chimpanzees and stated that "they sprawled out along comfortable branches ...or in day beds." At night they slept in arboreal nests on their sides or supine (1968, pp. 198-199). Goodall (1965, pp. 448-449) concluded that reclining on one side with hindlimb flexure was the "most normal" sleeping posture among her subjects and that they also occasionally slept prone. Izawa and Itani (1966, p. 147) concluded that chimpanzees generally slept supine though some slept on their sides.

CURRENT MODELS OF HOMINID POSTURAL EVOLUTION

The Great Apes (Yerkes and Yerkes, 1929) was published during the great debates among Henry Fairfield Osborn, William King Gregory, Sir Arthur Keith, F. Wood Jones and other paleontologists and anatomists about the role that brachiation might have played in predisposing humankind's ape-like ancestors to bipedalism. The problem of brachiation remains unresolved today despite frequent attacks upon it by behavioralists, experimental functional morphologists, evolutionary biochemists, paleoanthropologists, and quantitative comparative anatomists (Tuttle, 1975).

Currently there are four models on the evolution of hominid bipedalism. They may be designated (1) brachiating troglodytian, (2) knuckle-walking troglodytian (3) hylobatian, and (4) orangutanian. All four models are rooted to some extent in the classic brachiationist arguments of Keith and Gregory. But there has been a major shift away from referring to brachiation with a narrow connotation of gibbon-like rapid arm-swinging to concepts of suspensory behavior (Carpenter and Durham, 1969; Carpenter, 1976) which encompass a wide range of positions and activities wherein the forelimbs are upraised and subjected to tensile stresses (Tuttle, 1974, 1975).

The brachiating troglodytian model of Keith (1903, 1923, 1927, 1934) and Gregory (1927a-d) is premised chiefly on the observations of comparative anatomists which evidence close similarities between chimpanzees and man, especially in structure of the trunk and upper limb. The relatively large size of ancient troglodytians is supposed to have forced them to the ground where arboreally evolved predispositions to

orthograde posture served as a substrate for the transformation of their descendents into bipeds (Tuttle, 1974).

The knuckle-walking troglodytian model was also inspired
by the remarkable structural similarities between chimpanzees
and man. In addition, the extremely close biochemical features
in man and African apes led the author of the knuckle-walking
troglodytian model to conclude that they had had a very recent
common ancestry (Washburn, 1967, 1968, 1974). Washburn proposed that knuckle-walking provided a partial solution to
Keith's conundrum (1903, p. 19) about how novice terrestrial
apes could have survived and developed full-fledged bipedalism.
Knuckle-walking would permit the semi-erect troglodytians to
move rapidly away from predators. Because the fingers of
knuckle-walkers were flexed they could carry tools during
locomotion.

The hylobatian model of Morton (1924, 1926, 1927; Morton
and Fuller, 1952) fundamentally differed from the Keith-
Gregory brachiationist model in the body size of hypothetical
ante-hominid apes. Morton (1924) described them as about the
size of modern gibbons but stockier in build, with forelimbs
and hindlimbs of approximately equal length and well developed
grasping thumbs and big toes. They climbed versatilely in
trees, freely brachiated like chimpanzees (instead of rico-
chetally like gibbons) and habitually engaged in gibbon-like
bipedalism in trees and on the ground.

The orangutanian model was suggested by Stern (1976).
The orangutanians were smaller than modern orangutans (presumably females as well as males though Stern is not specific
about this). Like modern orangutans, they commonly employed
their forelimbs for versatile climbing and suspensory behavior.
But their hindlimbs predisposed them to more pronograde quadrupedalism than is characteristic of _Pongo_ (Stern, 1976,
p. 67). Unlike ancestral African apes, when the orangutanians
ventured to the ground their extremely elongate forelimbs precluded secondary adaptations to terrestrial quadrupedalism.
Their "only reasonable choice" was to free the forelimbs entirely from locomotor functions so they evolved into bipeds
(Stern, 1976, p. 68).

DISCUSSION WITH A VIEW TOWARD THE FUTURE

Despite more than a century of research, including
especially intensive and diversified studies during the past
15 years, we still cannot answer such basic questions as What
did the proximate ancestors of emergent Hominidae look like?
and How did they live? It is impossible to arrange the four
evolutionary models sketched hereinabove in a probabilistic
series upon which informed evolutionary anthropologists

would agree.

Peering into test tubes, even crystal ones, will not provide answers to questions about the morphological appearance and lifeways of pre- and protohominids. Only paleontological research in the interdisciplinary context of thorough-going experimental, functional morphological and ecological studies on extant primates can render major clues to the puzzle of hominid origins.

It should be clear from my survey of naturalistic studies on living apes that we are far short of comprehensive knowledge about their positional behavior. Unlike the fortuitous enterprise of fossil hominoid studies in which natural destructive processes have eliminated major bits of vital information, the opportunity still remains to round out our knowledge of positional and feeding behavior in Pan troglodytes, Pan paniscus, Pongo pygmaeus, and additional populations of Hylobates and Symphalangus in relatively undisturbed habitats. This possibility is vanishing rapidly not only because of commercial exploitation of natural forests but also because personnel at most scientific field stations provision pongid subjects or, in the instance of some orangutans, hand rear and release them into the forest.

Perhaps no greater tribute could be rendered to the memory of Dr. Yerkes than to fill the great gaps which he and Mrs. Yerkes astutely noted in The Great Apes and to augment the field studies of apes which he foresaw as essential to all aspects of modern primatology.

ACKNOWLEDGEMENTS

My career as a primatologist and evolutionary anthropologist would not have developed as it did without the Yerkes Regional Primate Research Center. I especially thank Professor Sherwood Washburn for sending me some 15 years ago to conduct thesis research at the picturesque facility in Orange Park, Florida, and Drs. Charles M. Rogers, Emil Menzel, Irwin Bernstein, and Richard Davenport for their hospitality and colleagueship there. Dr. Geoffrey Bourne and his staff have encouraged and accommodated many subsequent projects at the new facility in Atlanta, Georgia

This paper was written with support from NSF Grant No. SOC75-02478.

REFERENCES

Albrecht, H. and Dunnett, S. C. (1971). Chimpanzees in Western
     Africa. R. Piper and Co., Munchen.
Bingham, H. C. (1932). Carnegie Inst., Wash. Publ. 426: 1-66.
Bournonville, D. de (1967). Bulletin de l'Institut-Fondamental
     d'Afrique Noire 29(ser.A): 1188-1269.
Carpenter, C. R. (1934). Comp. Psych. Monogr. 10(2): 1-168.
Carpenter, C. R. (1935). J. Mammal. 16: 171-180.
Carpenter, C. R. (1938). Netherlands Committee for International
     Nature Protection, Amsterdam, Communications No. 12,
     pp. 1-34.
Carpenter, C. R. (1940). Comp. Psych. Monogr. 16(s): 1-212.
Carpenter, C. R.(1976). Gibbon and Siamang 4: 1-20, Karger, Basel.
Carpenter, C. R. and Durham, N. M. (1969). In "Recent Advan-
     ces in Primatology" (H. Hofer, ed.) Vol. 2, pp. 147-154.
     Karger, Basel.
Chivers, D. J. (1972). Gibbon and Siamang 1: 103-135. Karger,
     Basel.
Chivers, D. J. (1974). Contributions to Primatology 4: 1-335.
     Karger, Basel.
Davenport, R. K. (1967). Folia primat. 5: 247-263.
Ellefson, J. O. (1968). In "Primates" (P.C. Jay, ed.) pp. 180-
     199. Holt, Rinehart and Winston, New York.
Ellefson, J. O. (1974). Gibbon and Siamang 3: 1-136, Karger,
     Basel.
Fleagle, J. G. (1976). Folia primat. 26 (4): 245-269.
Galdikas-Brindamoor (1975). Natl. Geog. 148 (4): 444-473.
Goodall, J. M. (1962). Ann. N. Y. Acad. Sci. 102: 455-467.
Goodall, J. (1963a) Symp. Zool. Soc. London 10: 39-47.
Goodall, J. (1963b) Natl. Geog. 124: 271-308.
Goodall, J. (1965). In "Primate Behavior" (I. DeVore, ed.)
     pp. 425-473. Holt, Rinehart & Winston, New York.
Gregory, W. K. (1927a) Science 65: 601-605.
Gregory, W. K. (1927b) Quart. Rev. Biol. 2: 549-560.
Gregory, W. K. (1927c) Proc. Amer. Phil. Soc. 66: 439-463.
Gregory, W. K. (1927d). Sci. Amer. 137: 230-232.
Harrisson, B. (1962). Orang-utan. Collins, London.
Hornaday, W. T. (1879). Proc. Amer. Assoc. Adv. Sci. 28:
     438-455.
Hornaday, W. T. (1885). Two Years in the Jungle. Chas.
     Scribner and Sons, New York.
Horr, D. A. (1975). Primate Behavior, 4: 307-323. Academic
     Press, New York.
Itani, J. and Suzuki, A. (1967). Primates 8: 355-381.
Izawa, K. (1970). Primates 11: 1-46.
Izawa, K. and Itani, J. (1966). Kyoto Univ. Af. Stud. 1:
     73-156.

Jones, C. and Sabater Pi, J. (1971). Bibliotheca Primat.
    13: 1-96.
Kano, T. (1972). Kyoto Univ. Af. Stud. 7: 37-129.
Keith, A. (1903). J. Anat. Physiol. 37: 18-40.
Keith, A. (1923). Brit. Med. J. 1: 451-454, 499-502, 545-548,
    587-590, 624-626, 699-672.
Keith, A. (1927). Concerning Man's Origin. Watts, London.
Keith, A. (1934). The Construction of Man's Family Tree.
    Watts, London.
Kortlandt, A. (1962). Sci. Amer. 206(5): 128-138.
Kortlandt, A. (1967). In "Neue Ergebnisse der Primatologie."
    (D. Starck, R. Schneider and H.-J. Kuhn, eds.) pp. 208-
    224. Fischer, Stuttgart.
Kortlandt, A. (1968). In "Handgebrauch und Verständigung bei
    Affen und Fruhmenschen" (B. Rensch, ed.). pp. 59-102.
    H. Huber, Bern-Stuttgart.
Kortlandt, A. (1972). New Perspectives on Ape and Human Evolu-
    tion. Stichting voor Psychobiologie, Amsterdam.
Kortlandt, A. and Kooij, M. (1963). Symp. Zool. Soc., London.
    10: 61-88.
Lawick-Goodall, J. van (1968). Anim. Behav. Monogr. 1(3):
    161-311.
MacKinnon, J. (1971). Oryx 11(2-3): 141-191.
MacKinnon, J. (1973). Oryx 12(2): 234-242.
MacKinnon, J. (1974a). Anim. Behav. 22: 3-74.
MacKinnon, J. (1974b). In Search of the Red Apes. Collins,
    London.
McClure, H. E. (1964). Primates 5(3-4): 39-58.
Miller, G. S. Jr. (1942). Proc. Acad. Nat. Sci. Philadelphia,
    94: 107-165.
Morton, D. J. (1924). Am. J. Phys. Anthrop. 7: 1-52.
Morton, D. J. (1926). J. Morph. Physiol. 43: 147-179.
Morton, D. J. (1927). Am. J. Phys. Anthrop. 10: 173-203.
Morton, D. J. and Fuller, D. D. (1952). Human locomotion and
    body form. William & Wilkins, Baltimore.
Nissen, H. W. (1931). Comp. Psychol. Monogr. 8(1): 1-122.
Papaioannau, J. (1973). Malay. Nat. J. 26: 46-52.
Pocock, R. I. (1905) Proc. Zool. Soc. London 2: 169-180.
Prost, J. H. (1965). Amer. Anthrop. 67(5): 1198-1214.
Rahm, U. (1967). In "Neue Ergebnisse der Primatologie". (D.
    Starck, R. Schneider and H. J. Kuhn, eds.). pp. 195-207.
    Fischer, Stuttgart.
Rijksen, H. D. (1975). In "Contemporary Primatology" (S. Kondo,
    M. Kawai, and A. Ehara, eds.) pp. 373-379.
Rijksen, H. D. and Rijksen-Graatsma, A. G. (1975). Oryx 13(1):
    63-73.
Reynolds, V. (1965). Budongo, a Forest and Its Chimpanzees.
    Methuen and Co. Ltd., London.

Reynolds, V. and Reynolds, F. (1965). In "Primate Behavior"
(I. DeVore, ed.) pp. 368-424. Holt, Rinehart and Winston,
New York.
Rodman, P. S. (1973). In "Comparative Ecology and Behaviour
of Primates". (R.P. Michael and J. H. Crook, eds.) pp.
172-209, Academic Press, London.
Rose, M. D. (1973). Primates 14(4): 337-357.
Schaller, G. B. (1961). Zoologica 46(2): 73-82.
Schlegal, H. and Müller, S. (1839-1844). Zoologie 2: 1-28,
Leide.
Silva, G. S. de (1971). Malay. Nat. J. 24: 50-77.
Stern, J. T. Jr. (1976). Yearb. Phys. Anthrop. 1975. 19: 59-
68.
Sugiyama, Y. (1968). Primates 9: 225-258.
Sugiyama, Y. (1969). Primates 10: 197-225.
Sugiyama, Y. (1973). In "Behavioral Regulators of Behavior
In Primates" (C.R. Carpenter, ed.) pp. 68-80, Bucknell
University Press, Lewisburg, Pa.
Suzuki, A. (1969). Primates 10: 103-148.
Suzuki, A. (1971). J. Anthrop. Soc. Nippon 79: 30-48.
Tenaza, R. R. and Hamilton, W. J. (1971). Fol. Primat. 15:
201-211.
Tuttle, R. H. (1967). Amer. J. Phys. Anthrop. 26(2): 171-206.
Tuttle, R. H. (1974). Current Anthrop. 15: 389-426.
Tuttle, R. H. (1975). In "Phylogeny of the Primates."
(W.P. Luckett and F.S. Szalay, eds.) pp. 447-480.
Plenum, New York.
Wallace, A. R. (1869). The Malay Archipeligo. (1962 Dover
edition). Dover Publications, Inc. New York.
Washburn, S. L. (1967). Proc. Roy. Anthrop. Inst. Gr. Brit.
Ireland, pp. 21-27.
Washburn, S. L. (1968). The Study of Human Evolution (Condon
Lectures). Univ. of Oregon Books, Eugene.
Washburn, S. L. (1974). Yearb. Phys. Anthro. 1973, 17: 67-70.
Yerkes, R. M. and Yerkes, A. W. (1929). The Great Apes.
Yale University Press, New Haven.

Index

A
B 7
C 8
D 9
E 0
F 1
G 2
H 3
I 4
J 5